The 100th Anniversary of
China's Water Conservancy Collection

(VOLUME II)

100 GREATEST PROJECTS

Edited by Wu Zhongru, Tang Hongwu, Xu Hui, and Zhang Bing

Translated by Jianning Ding and Xuemeng Angela Li

Books Beyond Boundaries

ROYAL COLLINS

The 100th Anniversary of China's Water Conservancy Collection (Volume II):
100 Greatest Projects

Edited by Wu Zhongru, Tang Hongwu, Xu Hui, and Zhang Bing
Translated by Jianning Ding and Xuemeng Angela Li

First published in 2023 by Royal Collins Publishing Group Inc.
Groupe Publication Royal Collins Inc.
BKM Royalcollins Publishers Private Limited

Headquarters: 550-555 boul. René-Lévesque O Montréal (Québec) H2Z1B1 Canada
India office: 805 Hemkunt House, 8th Floor, Rajendra Place, New Delhi 110008

Original Edition © Hohai University Press

ISBN: 978-1-4878-1099-3

To find out more about our publications, please visit www.royalcollins.com.

THE 100TH ANNIVERSARY OF
CHINA'S WATER CONSERVANCY COLLECTION

Editor-in-Chief: Wu Zhongru

Tang Hongwu

Xu Hui

EDITORIAL BOARD

CONTENTS

FOREWORD

One hundred years is just a short moment in the course of time, but it is also a long journey in the history of human society. More than 4,000 years ago, Emperor Yu started the civilizational history of the Chinese people prosper the country with water regulation. Since then, successive generations of Yu's descendants have been working to develop water conservancy and eliminate floods to inherit and promote the undertakings, which left enormous stories of water regulation and set up countless monuments for water regulation. Names including Emperor Yu, Sunshu Ao, Ximen Bao, and Li Bing were as brilliant as stars, shining brightly in the sky of China. Famous water conservancy projects such as Anfengtang Reservoir, Zhengguo Canal, Yinzhang Shi'er Canals, Dujiangyan, and the Grand Canal are not only recorded in the glorious history of Chinese civilization but are also still benefiting the world after thousands of years.

As the wheel of history entered the modern era, a group of hydraulic professionals such as Li Yizhi, Zheng Zhaojing, Wang Huzhen, and Zhang Hanying inherited the ancient Chinese water regulation experience, on the basis of which they actively absorbed the technological achievements of modern civilization and opened a new chapter of modern Chinese water conservancy. After the founding of the People's Republic of China, as the foundation of the national economy, water conservancy witnessed rapid development and brilliant achievements under the leadership of the Party and the State. In addition to the radical cure of floods in major rivers, a large number of mega projects emerged from the Gezhouba Dam, Three Gorges Project, to the Yellow River Xiaolangdi Dam, etc. These projects transformed natural disasters into water resources and created comprehensive practical benefits through flood control, irrigation, power generation, navigation, and tourism.

The calmness of rivers that can be witnessed by the world and recorded in history has long been a long-cherished dream of the Chinese people, and it can only be realized under the leadership of the Communist Party of China. 2021 marked the 100th anniversary of the founding of the Communist Party of China. The development of modern Chinese water conservancy has been going on for more than 100 years. In order to reflect on the past, follow up on past masters' wisdom, inherit the Chinese civilization of water regulation, and promote the spirit of water conservancy and the spirit of China, Hohai University organized and planned the *100th Anniversary of China's Water Conservancy Collection*.

The collection includes two volumes: *100 Greatest Projects* and *100 Greatest Scientists*.

100 Greatest Projects selected 100 typical hydraulic projects of China and provided readers with a comprehensive understanding of the century-long arduous development course of China's hydraulic projects and their glorious achievements through the introduction of the development, construction management, and economic and social benefits of the projects. The volume also took into account the uniqueness of different hydraulic projects and selected different types of projects, including hydropower projects, water diversion projects, flood diversion projects, large-scale irrigation districts, dive projects, and port projects. It also introduced a large number of typical projects with their own characteristics, such as the Jinping-I Dam, Three Gorges, Longtan Hydropower Station, and other world's largest hydroelectric dams, including the Xin'anjiang Hydropower Station, the first large hydropower station in China that was surveyed, designed and constructed domestically.

100 Greatest Scientists was mainly composed of the glorious achievements of some academicians and experts (born before 1949) in the fields of water conservancy, hydropower, and water transportation. The book introduced the readers to the hydraulic scientists of modern China over the past hundred years, who studied both in the East and the West world and obtained achievements connecting the past and the present. Following the spirit of Emperor Yu and having the magnificent wish to regulate and develop rivers, they built dams, power stations, ports, and docks with efforts and wisdom. They built irrigation areas, dredged rivers and canals, and obtained extraordinary achievements in various aspects, including water conservancy, hydropower, water transportation, etc. They held the belief that "I do not have to be credited to success but must contribute my parts to it." They have acted with dedication and selflessness. They had always been chasing after the brightness in their heart regardless of the difficulties with the fire of their ideal of fighting for the water conservancy undertakings constantly burning.

Reflect on the past, and look forward to the future. Regulating water to benefit people has been a significant undertaking for thousands of years. The majestic dams are the witness to the prosperity of the world. Water is always closely related to people's livelihood. It is hydraulic professionals' and departments' responsibility and mission to consolidate the foundation of water security, enhance people's well-being, and constantly improve people's sense of achievement, happiness, and security.

Currently, the situation of China's water security is still serious. Hydraulic professionals must implement General Secretary Xi Jinping's new ideas, thinking, and requirements on water regulation and prosperity, which is to "prioritize water conservation, balance the space, regulate systematically, and utilize both the state and the market's functions." They must coordinate the work between water disaster prevention and control, water conservation, water ecological protection and restoration, water environment control, etc., to constantly accelerate the upgrading of watershed protection and governance system, improve the flood prevention and mitigation system, comprehensive utilization of water resources system, water ecological and environmental protection system, comprehensive management system of watershed, and spare no efforts to protect national water security.

Over the past hundred years, the development of China's water conservancy undertakings has witnessed the course of the Communist Party of China leading the Chinese people to prosperity and prosperity. It recorded how the great national projects protected the peace of rivers, advanced economic development, and created a better life. During this century, hydraulic professionals endured difficulties, only to focus on national rejuvenation. During this century, hydraulic professionals never forgot their original intention and were only devoted to the well-being of the people. All their names will shine forever in the history of China's water conservancy construction and development, representing the spirit of water conservancy and national pillar. All these majestic water conservancy projects, standing magnificently in the rivers, are symbols of national power and people's wisdom, highlighting the spirit and confidence of China.

The spirit of Emperor Yu includes a sense of charity, the spirit of fighting, the pursuit of innovation, a scientific attitude, and open-mindedness. The lives and deeds of generations of outstanding hydraulic professionals are the most vivid portrayal of the "Spirit of Emperor Yu." All the major hydraulic projects are eternal monuments of China's practice of water regulation following Emperor Yu, and will always be a source of strength and models for future generations.

The 100th Anniversary of China's Water Conservancy Collection is a beneficial attempt to tell the story of Chinese hydraulic professionals, to promote the spirit of water conservancy, the spirit of China, and to show the power and confidence of China. It is not only suitable for the theme of the era but also of realistic significance. We hope that this series will be seeds that continue to sprout, blossom, and bear fruit until eventually become a great spectacle.

EDITORIAL GROUP
February 2021

JINPING-I HYDROPOWER STATION
The World's Highest Double-Curvature Arch Dam

Jinping-I Hydropower Station is located in Muli County and Yanyuan County, Liangshan Yi Autonomous Prefecture, Sichuan Province. It is a controlled reservoir cascade power station in the lower reaches of the main stream of the Yalong River, about 358 km away from the estuary, and 75 km directly from Xichang City. It is the leading one among the five cascade power stations planned from Kala to Jiangkou in the lower reaches of the Yalong River main stream. The basin area above the dam site is 103,000 km², accounting for 75.4% of the Yalong River basin. The multi-year average flow is 1,220 m³/s, and the average annual runoff is 38.5 billion m³. The project is mainly for power generation, water storage in flood season, and sharing the task of flood control in the middle and lower reaches of the Yangtze River.

Jinping-I Hydropower Station panorama

With the world's highest dam of 305 m, the engineering geology condition of Jinping-I Hydropower Station is extremely complicated. It is featured with "5 Highs and 1 Deep," that is, high mountain canyon, high arch dam, high slope, high in-situ stress, high head, and deep unloading. It is acknowledged by experts at home and abroad to be a mega hydropower project which has "the most complicated geological conditions, the worst construction environment" and is "technically and managerially most difficult." Preliminary survey and design work started in the 1950s. The feasibility study was approved in November 2003. Construction was officially approved in September 2005. The river interception was on December 4, 2006, and the first two units were put into operation to generate electricity on August 30, 2013. The completion of the power station was in July 2014, and all units were put into operation. Upon completion, the power station will supply power to Sichuan, Chongqing, and East China, with 50% going to East China.

The project is huge in scale and mainly consists of a concrete double-curved arch dam (including cushion pond and second dam), a right bank spillway tunnel, a right bank diversion power generation system, and a switching station. The total storage capacity of the reservoir is 7.765 billion m³, the normal storage level is 1,880 m, the dead water level is 1,800 m, the storage capacity below the normal storage level is 7.76 billion m³, and the adjusted storage capacity is 4.91 billion m³, acting as an annual regulated reservoir. The total installed capacity of the power station is 3,600 MW, and the multi-year average power generation capacity is 16.62 billion kW·h. The annual utilization hours are 4,616 h.

During the construction, it has set a number of world records, including: ① The world's highest double-curvature arch dam with a height of 305 m; ② The world's largest and most complicated extra-high arch dam foundation treatment project; ③ Treatment of high and steep slopes with the world's most complex geological condition at dam abutment; ④ The first attempt and the world's largest in range and scale in the treatment of dangerous rock mass of the natural slope in the key area of hydropower project; ⑤ The general constructive layout of the extra-high arch dam was successfully carried out in the alpine and canyon area, and for the first time, the layout problems of traffic, water supply, power supply, plant facilities, material yard and slag dump site, and construction camp were systematically solved; ⑥ The world's first super-high arch dam with multi-layer orifice collision-free energy dissipation; ⑦ The world's first spillway tunnel in the form of pressurized jointed unpressurized dragon drop tail using the dovetail can pick flow energy dissipation worker; ⑧ The world's highest independent shore tower power station inlet with stratified water intake function, which provides ecological and environmental protection; ⑨ The world's first large-scale underground cavern cluster project successfully completed under high in-situ stress environment and low rock strength ratio; ⑩ The world's largest cylindrical impedance-type tailwater regulator chamber, with a diameter of 41.0 m and a height of 80.5 m.

On November 29, 2014, a ceremony was held to celebrate the construction completion and full operation of Jinping-I Hydropower Station, the mega hydropower station in Sichuan Province, with the world's highest arch dam and the world's largest hydraulic tunnel group. 14 units of 600,000 kW have been fully completed and put into operation, and the installed hydropower capacity of Sichuan Province has exceeded 60 million kW, accounting for more than 80% of the total installed power capacity, ranking first in China in both quantity and quality. With the normal power generation of Jinping-I Hydropower Station, 7.682 million tons of raw coal will be saved annually, reducing the emissions of sulfur dioxide 105,000 t, of carbon dioxide by 13.712 million. It promotes energy conservation and emissions reduction and is vital to clean energy development. It also increases the average output of the Sichuan power grid by 22.5% during the dry period, which greatly optimizes the power supply structure of the Sichuan and Chongqing power grids. Each year, the Yalong River downstream cascade power station increases its power generation capacity by 6 billion kW·h, and the Xiluodu, Xiangjiaba, Three Gorges and Gezhouba hydropower stations in the Jinsha River increase the power generation by 3.77 billion kW·h.

The double-curvature arch dam with the highest water retaining first adopts digital means for the intelligent building of hydropower projects. Current first-class construction techniques in the field of hydropower infrastructure engineering, such as unmanned crushing, digital command systems, concrete pouring, and high-slope intelligent maintenance systems, laid the foundations for the world's first smart dam. The hydropower industry has a saying, "Three Gorges is the largest project, but Jinping is the most complex one." Construction difficulty is rare in the world. It is home to the world's tallest double-curved arch dam. By taking advantage of the natural drop of the Yalong River in Jinping Bay, the cutting bend is straightened to excavate the tunnel to divert water to generate electricity. And thus, the world's largest mountain tunnel was successfully opened. The project has stood flood tests many times since it was impounded to the normal water level in August 2014. The deformation, stress, seepage, and osmotic pressure of the dam, foundation, and high slope all meet the design requirements and operate in good condition. The project has made outstanding technological innovation achievements in the deformation control of the complex foundation of the extra-high arch dam, the stability and safety control of the high and steep slope, the stability of the surrounding rock of the underground powerhouse cavern group with high in-situ stress and structural development, the high head, the ultra-high velocity, the large volume of flood discharge, the anti-cracking of the extra-high arch dam concrete, and the remarkable social economic and environmental benefits.

On June 3, 2018, the project won the 15th China Civil Engineering Zhan Tianyou Award, the 2015 World Federation of Engineering Organizations Award for Outstanding Engineering Construction, the 2018 FIDIC Engineering Project Outstanding Achievement Award, known as the "Nobel Prize" in engineering consulting, awarded by the International Federation of Consulting Engineers. In 2018, the Jinping-I project won the Outstanding Achievement Award. Only two projects in the world were awarded, and Jinping was in the first place, highly recognized by the international engineering community.

Jinping-I Hydropower Station, as the high-quality resource from the Yalong River main stream cascade development, will generate huge comprehensive benefits such as power generation, flood control, sand control, shipping, energy conservation, and emission reduction, and will greatly help local poverty alleviation. Its development and construction can drive local economic development and is of profound significance to promoting the complementary advantages of western resources and eastern economic advantages.

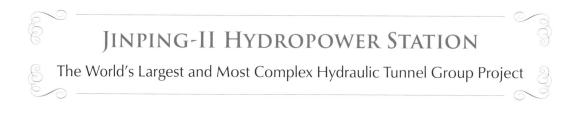

Jinping-II Hydropower Station

The World's Largest and Most Complex Hydraulic Tunnel Group Project

Jinping-II Hydropower Station is located on the main stream of the Yalong River in the territory of Liangshan Yi Autonomous Prefecture, Sichuan Province. It is the backbone power station of the cascade development in the lower reaches of the Yalong River, among the hydropower stations with the highest water head, the largest power generation, and the best benefit. The project was officially started on January 30, 2007, put into operation for power generation in December 2012, and the hub project passed the completion inspection in January 2016. The total investment is 38.056 billion yuan.

Jinping-II Hydropower Station Dam

Jinping-II Hydropower Station takes advantage of the natural drop of the 150 km large river bay of the Yalong River, and diverts the main stream through the 16.67 km long diversion tunnel to a straight line for power generation, creating an effective "big river bay development model." The hub of Jinping-II Hydropower Station is mainly composed of three parts: the first barrage sluice dam, the diversion system, and the underground powerhouse. It is a diversion power station with a low gate, long tunnel, high head, and large capacity. The dam is located 7.5 km downstream of the Jinping stage I project, and is a large (I)

Jinping-II Hydropower Station Gate Dam

type project with a rated head of 288.0 m. Eight sets of 600 MW hydro-generating units are installed, with an installed capacity of 4,800 MW and an average annual power generation capacity of 24.23 billion kW·h. The first barrage dam is located at the western end of the Yalong River. The first sluice dam is located at Maomaotan at the west end of the Yalong River Bay, with a maximum dam height of 34 m; the underground powerhouse is located at Dashuigou at the east end of the Yalong River Bay, adopting the tail development method; the water diversion system consists of the power station inlet, the water diversion tunnel, upstream regulator room, high-pressure pipeline, tailgate room, tailwater tunnel, and tailwater outlet building. The water inlet of the power station is located at Jingfeng Bridge, about 2.9 km upstream of the gate site, and the first hub adopts a layout plan in which the gate and dam are separated from the water inlet. The watershed above the gate site covers an area of 0.103 million km², and the average multi-year flow rate at the gate site is 1,220 m³/s. The normal storage level of the reservoir is 1,646 m, with a total storage capacity of 14.28 million m³. The adjustment storage capacity is 5.02 million m³, which itself has daily adjustment performance, and has the same annual adjustment characteristics as the first-level synchronous operation of Jinping.

The Jinping-II Hydropower Station is the hydropower station with the highest water head, the largest power generation, and the best benefit among the cascade hydropower stations on the Yalong River. It is also a landmark project for the strategic implementation of "West-East Power Transmission" and "Sichuan Power Transmission." The large-scale deep-buried tunnel group consists of four diversion tunnels, two auxiliary tunnels, and one construction drainage tunnel, with a total length of 120 km, featuring deep burial, long tunnel line, big tunnel diameter, high in-situ stress level, complex karst hydrogeological conditions, and difficult construction layout, etc. It is the world's largest and most difficult hydraulic tunnel group project ever built and is under construction. During the construction of the project, it was faced with world-class technical problems such as the excavation of 2,500 m-level ultra-deep tunnels, the damage of strong rock bursts, and the hazard treatment of ultra-high pressure inrush water with a kilometer-level head. Through technical innovation, it has created the world record of 3,300 m of tunnel group excavation monthly, and 58 months of complete penetration, and created the "Quantitative Identification of Unfavorable Geology in Advance Like Water-bearing Structures in Tunnels and its Key Technology for Disaster Prevention and Control," "Key Technology for Mechanism, Early Warning and Dynamic Control of Hard Rock High-Stress Disaster Breeding Process," "Key Technology and Engineering Application for Major Surge Water Disaster Management in Tunnels and Underground Engineering." "Key Technology for Power Generation Project of Jinping-II Super-deeply Buried Super-large Diversion Tunnel," etc. Many world-class technical problems were successfully overcome, such as high burial depth, high ground stress, high-pressure water inrush, high water head, and large-capacity unit design and manufacturing, and thus the project won the Second Prize of National Science and Technology

Jinping-II Hydropower Station underground powerhouse

Progress. The excellent quality evaluation of the Jinping-II Hydropower Station project is of milestone significance for the construction of China's ultra-long diversion tunnel and for leading the country's hydropower scientific and technological progress. In 2020, it won the Zhan Tianyou Award for Civil Engineering in China.

Jinping-II underground cave complex

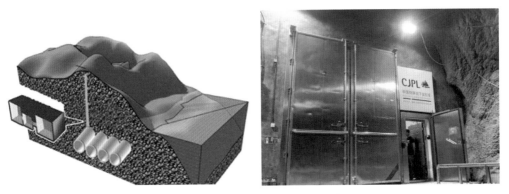
Jinping Extremely Deep Underground Dark Matter Laboratory

In addition, China's first extremely deep underground laboratory, "China Jinping Underground Laboratory," built by using the Jinping Mountain tunnel built for the hydropower station, was unveiled and put into operation on December 12, 2010, with a vertical rock coverage depth of 2,400 m. At present, the dark matter laboratory has the deepest rock coverage, the largest volume, the smallest cosmic ray flux, and the complete infrastructure in the world. "Dark matter" is a kind of cosmic material that is extremely difficult to detect. In order to detect dark matter, the laboratory needs to be built under the most demanding natural conditions in the mountains. Only in this way can the interference of cosmic rays be shielded. The completion of the laboratory marks that China has a world-class clean low-radiation research platform and can independently carry out the most cutting-edge international basic research topics, such as dark matter detection. The dark matter detector of the Tsinghua University experimental group took the lead in entering the laboratory and started the detection work; Shanghai Jiao Tong University and other research teams also entered here one after another to carry out dark matter detection research. This is of great significance to promoting independent research and applied research on major fundamental frontier topics in China.

LONGTAN HYDROPOWER STATION
The World's Highest Compact Concrete Dam

Longtan Hydropower Station is located on the Hongshui River in Tian'e County, Guangxi Zhuang Autonomous Region, and is the main backbone project in the development of the Hongshui River, the landmark project of "West-East Electricity Transmission," and also the key project of the Western Development. The main project of Longtan Hydropower Station started construction on July 1, 2001, with a total period of 9 years. The river was cut off in 2003, the first unit generated electricity in June 2007, and all units were put into power generation in 2009.

Longtan Hydropower Station panorama

Longtan project is mainly composed of a compact concrete gravity dam, power plant, water release building, navigation building, and so on. The dam is a compact concrete gravity dam, and the dam axes on both banks are turned upstream by 27° and 30°, respectively, due to the influence of the topography and geological conditions of the dam site, and are in a folding shape. Constructed in two phases, the maximum dam height is 192 m in the early stage. At the end of the second phase, the dam crest elevation is 406.5 m, the maximum dam height is 216.5 m, and the top length of the dam is 849 m. The normal storage level of the reservoir is 400 m, the total storage capacity is 27.27 billion m³, and the flood control storage capacity is 7 billion m³. Longtan has a huge underground plant with a total length of 388.5 m, a net width of 28.5 m, and a total height of 74.4 m. The plant is connected to a pressure diversion tunnel behind the powerhouse, with a dam-type inlet, a single machine and a single pipe diversion, and a tailwater regulating well and a tailwater tunnel for every three machines. The flood discharge building is arranged in the riverbed part of the dam, with seven table holes and two bottom holes. The bottom holes are arranged on both sides of the table holes; the table and bottom holes are used to pick up the flow and dissipate energy; the bottom holes are not responsible for flood discharge tasks, mainly used for late inflow, reservoir emptying, and sand flushing tasks; the navigation building is a two-stage counterweight vertical boat lift, with a total length of 1,700 m, the maximum lift of 179 m. It is lifted in two stages, 88.5 m and 90.5 m, respectively.

Longtan has the world's highest compact concrete dams, the largest underground plant, and the highest boat lift in the world. Fruitful results have been achieved in the design of highly compact concrete dams, anti-seepage, rapid construction in high temperature and rainy seasons, and dam-building materials. ① The use of compact concrete dam construction technology in the high seismic intensity zone to build a 200 m high compact concrete dam; ② The use of high-lift vertical boat lift to solve the hub navigation problems; ③ Adopting a combination of traditional methods and the latest modern research means to design the 400 m level left bank inlet high side slope; ④ The use of cross-sectional economic, self-

Longtan Compact Concrete Dam panorama

impermeable compact concrete gravity dams, reducing cement consumption by about 30%, thus energy-saving and environmental friendly; ⑤ The diversion power generation system is all arranged underground on the left bank, which greatly reduces the amount of open excavation works and the impact on the natural environment; ⑥ The project uses the excavated and abandoned slag to form a site of about 425,400 m², which saves the engineering investment; ⑦ The hydraulic turbine operating water head has a wide range of amplitude (97–179 m). In order to make the unit have better characteristics and unit stability under the condition of high operating water head difference, new research has been carried out in the selection of electromechanical parameters of hydro-generator sets, unit stability, and the electromechanical technology design research, equipment matching selection, unit manufacturing, and installation process, etc.; ⑧ The GPS comprehensive decision support system for Longtan Reservoir immigrants has been developed, which can get timely information of the environmental and socio-economic factors in the area around the power station and the general situation of land resources. The system has been developed to achieve scientific, systematic, and informationalized inundation treatment planning and design. The Longtan Hydropower Project is of excellent quality and has won the FIDIC Centennial Engineering Project Excellence Award. In 2007, after an inspection of the Longtan Hydropower Project, Mr. Louis Berger, Chairman of the International Commission on Dams, gave a very high evaluation—"The Longtan Dam sets a new record in the world for compacted concrete dams, with good dam construction quality and excellent performance."

The Longtan Hydropower Project is of great significance. As a backbone project in the flood control engineering system of the middle and lower reaches of the Northwest River and the Pearl River Delta, it greatly reduces the flood threat to 17 million people and 7 million mu (1 mu ≈ 666.7 m²) of arable land in the downstream area. After the completion of the project, it has canalized 224 km of waterway in the reservoir area, providing a convenient

Longtan Reservoir

and economic waterway to the sea for the inland province of Guizhou, and promoting the revitalization of the shipping business along the river and local economic development. Longtan Hydropower Station has huge power generation benefits and is a strategic project to improve the energy structure of Guangxi and even South China. The total installed capacity of the project is 6.3 million kW, with an average annual power generation capacity of 18.7 billion kW·h. After the completion of the project, more than 50% of the power will be sent to Guangdong, which will be included in the power balance as the power supply point of Guangdong during the "Eleventh Five-Year Plan" period. Obviously, the construction of Longtan Hydropower Station is of practical and strategic significance for promoting the eastward transmission of electricity from the west, promoting national networking, realizing the optimal allocation of energy resources in the country, meeting the needs of power growth in Guangdong and Guangxi, optimizing the power supply structure and power structure in southern China, reducing the flood threat in the lower reaches of the Hongshui River and the areas on both sides of the Xijiang River, and promoting the economic and social development of minority areas in Guangxi and Guizhou provinces. Meanwhile, since its operation, Longtan Hydropower Plant has also played an active role in flood control and irrigation, salinity suppression and replenishment in the Pearl River Estuary, soil and water conservation, water quality improvement, ecological balance maintenance, energy saving, and emission reduction, and energy storage compensation.

NUOZADU HYDROPOWER STATION PROJECT

The World's Third Highest and Asia's Highest Earth-Rock Dam

Nuozadu Hydropower Station is located downstream of Lancang River main stream at the junction of Cuiyun District and Lancang County, Pu'er City, Yunnan Province, and is the fifth stage of the eight-stage plan for the middle and lower reaches of Lancang River. Upstream and downstream are the constructed Dachaoshan Hydropower Station and Jinghong Hydropower Station, respectively. The power station is 215 km away from the upstream Dachaoshan Hydropower Station and 102 km away from the downstream Jinghong Hydropower Station. The power station is mainly for power generation, taking into account the flood control task in Jinghong city and farmland, and has comprehensive benefits such as improving shipping and developing tourism. The normal reservoir storage level is 812 m, the flood limit level is 804 m, the dead level is 765 m, the total reservoir capacity is 23.703 billion m^3, and the regulating reservoir capacity is 11.335 billion m^3, with multi-year regulation capability. The power station has installed nine units of 650,000 kW, with a total installed capacity of 5.85 million kW and a guaranteed output of 2.4 million kW. The multi-year average power generation is 23.912 billion kW·h, and can save 9.56 million tons of standard coal and reduce carbon dioxide emissions by 18.77 million tons. As the then

Nuozadu Hydropower Station panorama

largest hydropower station in Yunnan Province upon its completion, Nuozadu Hydropower Station is the backbone project for achieving the national goal of optimal resource allocation and national networking, and is the foundation project for the implementation of the "West-East Power Transmission" and "Cloud Power Transmission" strategies.

The main project of Nuozadu Hydropower Station started in January 2006, the engineering interception began in November 2007, the dam body began to be filled in January 2008, the diversion hole was launched in November 2011, the first unit was put into operation in July 2012, the dam project was completed in June 2013, and the reservoir was stored to the normal storage level of 812 m; as of May 28, 2014, all nine units have been put into operation. The construction was completed in June 2015, and the main project took 8.5 years to complete, two years earlier than the scheduled construction period.

Nuozadu Hydropower Station is composed of a heart wall rockfill dam, open spillway, and water diversion power generation system. The main body is a gravel soil heart wall rockfill dam with a maximum height of 261.5 m, which is Asia's highest and the world's third-highest earth-rock dam. The total capacity of the reservoir is 23.703 billion m³, which is equivalent to the storage capacity of 16 Dianchi pools. The top of the dam is 824.1 m, the lowest elevation of the bottom of the core wall is 560 m, the upstream side of the dam is equipped with concrete wave walls,

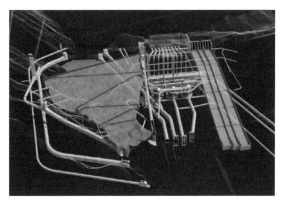

Layout of Nuozadu Hydropower Station

the top of the dam is 630.06 m long, the width of the top of the dam is 18 m, the slope gradient of the upstream dam slope is 1:19–1:1.833, the slope gradient of the downstream dam slope is 1:1.8–1:1.737. The open spillway is the largest in Asia, with the world's highest discharge power and flow rate. The spillway is arranged on a natural pass on the left bank, which greatly reduces the amount of excavation work. The horizontal length is 1,445.18 m, and the width is 151.5 m, including the inlet channel section, gate control section, drainage channel section, flip bucket section, outlet dissipation pond section, etc. The water diversion and power generation system is located on the left bank of the dam, with 87 main caverns, 26 construction branch caverns, and a total length of 22 km, with a total excavation volume of 2.9 million m³, a total concrete volume of 860,000 m³ and a metal structure and equipment installation volume of about 11,000 t.

The engineering characteristics of Nuozadu Hydropower Station are as follows.

(1) The total reservoir capacity is equivalent to 16 Dianchi lakes.

It is reported that after completing the task of water storage over several years, the total reservoir capacity of Nuozadu Power Station will reach 23.703 billion m³, equivalent to 16 Dianchi Lake. The reservoir of the power station has a multi-year regulation capacity, and after completion, the flood control standard of downstream Jinghong City can be increased from a 20-year return period to once in 100 years.

(2) The first domestic digital dam control.

Nuozadu Hydropower Station has six leading technologies in China: innovative application of digital dam management system, pioneering use of clay core wall gravel doping process, implementation of layered water extraction scheme, promotion of TOFD inspection technology for welded seams, completion of biodiversity protection facilities, and realization of "zero discharge" of production wastewater and domestic sewage, creating the eight typical achievements of project management: the best project management, the shortest construction period, the first regulation, outstanding scientific and technological innovation, friendly engineering environment, outstanding social responsibility, local development promotion, and peace and harmony in the work area. Since the first storage in 2013, the project has passed several tests during the flooding period and is in good operation. The maximum seepage volume of the project is 15 L/s, which is the smallest of similar projects at home and abroad. It can be seen that the project has excellent construction quality, good operation conditions, and outstanding technical innovation, and it is known as the world's most representative international milestone project of high rockfill dam, and won the China Civil Engineering Zhan Tianyou Award in 2018.

(3) **Create a green hydropower demonstration project.**

Nuozhadu Hydropower Station has made great efforts in ecological protection by building fish breeding stations, rare plant gardens, and animal rescue stations to save wild animals in need of protection before water storage, protecting the biodiversity of the Lancang River basin and achieving a complementary relationship between hydropower development and environmental protection.

Nuozadu Hydropower Station Observation Deck

The construction of Nuozadu Hydropower Station reflects the key requirements of the national implementation of the Western Development Strategy, which is not only an important national infrastructure facility, but can also drive the rapid development of industry, agriculture, tourism, fishery, and township enterprises and other industries in Yunnan Province, especially in Pu'er City, forming a new economic development belt in the Lancang River Basin, converting the development potential of the western region into real productivity, transforming potential markets into real markets, and turning resource advantages into economic advantages.

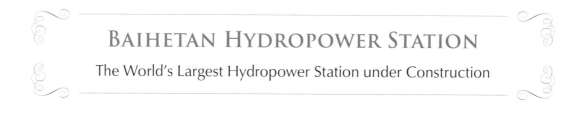

BAIHETAN HYDROPOWER STATION
The World's Largest Hydropower Station under Construction

The ancient people described the Jinsha River and the precipitous peaks beside it as such: Cliffs reach the sky and view the river like a well. But they could not have imagined that their descendants would have the courage to build a high dam side by side with the steep mountains. Located in Ningnan County, Sichuan Province, and Qiaojia County, Yunnan Province, Baihetan Hydropower Station is the second of the four cascades in the lower reaches of the Jinsha River—Wudongde, Baihetan, Xiluodu, and Xiangjiaba. Baihetan Hydropower Station is about 41 km upstream of Qiaojia County and 182 km from Wudongde Dam Site, about 195 km downstream from Xiluodu Hydropower Station, and about 380 km from Yibin Riverway. Baihetan Hydropower Station Project is a key project to carry out the important instruction of General Secretary Xi Jinping to "adhere to ecological priority and green development, promote the development of Jinsha River hydropower resources in a scientific and orderly manner" and implement the requirements of the Party Central Committee and the State Council to "promote the construction of ecological civilization, accelerate the adjustment of energy structure and supply-side reform, ensure energy security and ecological civilization construction," vigorously promote the development of hydropower in southwest China and increase the supply of clean electricity.

Baihetan Reservoir renderings

Baihetan Double-Curvature Arch Dam under construction

Baihetan Hydropower Station is a large (I) type project, currently the world's largest hydropower station under construction with the largest single-unit capacity. Hub project static investment of 143.07 billion yuan, dynamic investment of 177.89 billion yuan. The design and planning of the total construction period are 144 months, including 40 months of engineering preparation and 80 months of the main project construction. The construction is scheduled to be completed within two years. The project is composed of main buildings such as a barrage, flood discharge and energy dissipation facilities, and a water diversion power generation system.

Baihetan Reservoir controls a basin area of 430,300 km², with a total reservoir capacity of 20.627 billion m³, controlling 91% of the Jinsha River basin area, ranking first in China in terms of the capacity of high arch dams. It is an annual regulating reservoir with a regulating capacity of 10.436 billion m³ and a flood control capacity of 7.5 billion m³, second only to the Three Gorges Project and the Danjiangkou Project, with superb flood control capacity, and is an important part of the Yangtze River flood control system. The normal storage level is 825 m, and the flood control limit level is 785 m.

The barrage is a concrete double-curvature arch dam. The bottom elevation is 545 m, the top elevation is 834 m, the maximum dam height is 289 m, it is a 300 m high arch, the thickness of the dam arch is 14 m, the maximum arch end thickness is 83.91 m, including the maximum thickness of 95 m of the expanded foundation. The concrete (low-heat cement concrete) pouring volume is about 8.03 million m³.

The underground plant adopts the first part development method, and is the world's largest underground plant under construction. The left and right bank underground plants are 453 m long, 88.7 m high, and 34 m wide, which can fit into several 30-story skyscrapers, with a maximum horizontal depth of 1,050 m and a vertical depth of 330 m, a space equivalent to 1.5 "Water Cube" stadium size. It is the first hydropower project in the history of hydropower to adopt a single 1 million kW unit, the "Everest" in the world hydropower industry. The water diversion system adopts a single machine and a single cavern, with a single flow rate of 547.8 m³/s. The two machines of the tailwater system share a tailwater regulating chamber and a tailwater tunnel. A number of indicators of the cavern chamber excavation far exceeded the national standards, and all reached the exemplary quality level.

After more than ten years of scientific research, survey, design, and more than six years of careful preparation, the Baihetan Hydropower Station project entered the full construction phase of the main project on August 3, 2017. On April 6, 2021, the No. 2 diversion bottom hole of Baihetan Dam was lowered, and the water level in the reservoir area began to rise slowly, marking the start of the world's largest hydropower station under construction to lower the gate for water storage. It is expected that the first units of Baihetan Hydropower Station will be put into operation in July 2021, and all units will be put into operation in July 2022. The whole project will be completed in 2023.

Baihetan Dam Project

On the one hand, the Baihetan Hydropower Station project is known as one of the most complex hydropower projects in China and even in the world, and the main characteristic indexes are among the top in the world. It is the top in the world in terms of single-unit capacity (1 million kW), the size of cylindrical tailwater regulating wells, the size of underground cavern group, top in the world in terms of seismic parameters of 300 m class high dam, use of low-heat cement concrete for the whole dam of 300 m class high dam, and the high dam size of pressureless flood relief cavern group, etc. On the other hand, Baihetan Hydropower Station has huge comprehensive utilization benefits: ① Play the role of a green engine. After the hydropower station is completed and put into operation, the multi-year average annual power generation is 62.443 billion kW·h; the guaranteed output is 5,500 MW; the average annual reduction of standard coal consumption is about $1,986 \times 10^4$ t; the reduction of carbon dioxide emissions is about $5,160 \times 10^4$ t; the installed capacity becomes second only to the Three Gorges Project. The installed capacity will become the second largest hydropower plant in the world after the Three Gorges Project, which will promote the optimization of the national energy structure and the realization of non-fossil energy emission reduction commitments; ② Improve flood control standards. After the completion of the project, Wudongde reservoir and others' joint operation will improve Yibin, Luzhou, Chongqing, and other cities' flood control standards, realize joint scheduling with the Yangtze River Three Gorges reservoir, and raise the flood control standards in the middle and lower reaches of the Yangtze River; ③ Improve shipping conditions. More ships will be able to sail along the river. After the completion of the project, the reservoir shipping conditions have been greatly improved. Combined with the construction of highway and railroad traffic, the Jinsha River downstream comprehensive transportation system is formed, and thus will further enhance the Yangtze River "Golden Waterway" function. It is of great significance to the promotion of the economic and social development of the lower reaches of the Jinsha River region; ④ Help Sichuan and Yunnan provinces fight against poverty. The construction of Baihetan Hydropower Station can turn local hydropower resources into economic advantages, accelerate poverty alleviation in the reservoir area, and make a greater contribution to local economic and social development. In addition, as a rising "Great National Power," Baihetan Hydropower Station is also the backbone of China's "West-East Power Transmission" and the new "national card" of China's hydropower project besides the Three Gorges Project.

XIAOWAN HYDROPOWER STATION
The World's First 300 M-Class Double-Curvature Arch Dam

Xiaowan Hydropower Station is located 1.5 km downstream of the confluence of the Heihui River in the middle and lower reaches of Lancang River at the junction of Nanjian and Fengqing counties in Yunnan Province. It is the second stage of the "two reservoirs and eight stages" cascade development in the middle and lower reaches of Lancang River, with six single 700,000 kW units installed, a total capacity of 4.2 million kW, guaranteed annual power generation capacity of 19 billion kW·h, total reservoir capacity of 14.9 billion m³, normal reservoir storage level of 1,240 m, dead water level of 1,166 m. It is mainly for power generation, as well as flood control, irrigation, sand control, shipping, etc. It is the leading power station in the middle and lower reaches of the Lancang River.

The construction of Xiaowan Hydropower Station began in June 1999, and officially started on January 20, 2002, achieving the cut-off of the river one year ahead of schedule on October 25, 2004, the first concrete bin of the dam was poured on December 12, 2005, and the diversion hole was put into storage on December 16, 2008. The first unit was put into operation to generate electricity on September 25, 2009, and the goal of "three deliveries a year" was achieved the same year. All units were put into operation in 2010. On March 8, 2010, the entire line of the Xiaowan Hydropower Station dam

Xiaowan Hydropower Station panorama

was poured and capped. Thus, the world's tallest 300 m class double-curvature arch dam was officially born. The pivot project consists of a concrete double-curvature arch dam, a water pad pond and a second dam behind the dam, a dam body and a flood relief building on the left bank, an underground water diversion power generation system on the right bank, and a 500 kV ground switching station, etc. The peak horizontal acceleration of the bedrock, the length of the dam roof arc, the total water thrust, and other key indicators all rank first in the world arch dam construction.

Xiaowan Extra-high Arch Dam flood release

The construction of Xiaowan Hydropower Station overcame a number of technical problems, promoting China's arch dam design, construction, and other technologies to a new level.

(1) The dam is a parabolic-shaped variable thickness double-curvature arch dam with a crest elevation of 1,245 m and a maximum height of 294.5 m, the world's first 300 m double-curvature arch dam.

(2) The geological environment of the Xiaowan dam site area is complex, and earthquakes have been frequent throughout history. By the National Seismological Bureau Geological Research Institute's investigation and identification, the basic earthquake intensity of the Xiaowan hub area is VIII degrees, so seismic analysis needs to be done in accordance with 600 years beyond the probability of 10% when the peak horizontal acceleration is 0.308 g. Xiaowan's high arch dam seismic safety evaluation, stress control standards and engineering seismic measures, high arch dam shoulder static conditions, and stability analysis under dynamic conditions are the key technical problems in the world of dam design today. Therefore, the dam design and the research on seismic engineering measures of the Xiaowan project are the best of its kind in the world.

(3) The Xiaowan flood discharge and energy dissipation building consists of five table holes, six middle holes, water pad ponds, a second dam, and one flood discharge tunnel on the left bank

of the dam. The calibrated flood discharge volume is 20,683 m³/s, and the corresponding flood discharge power is 46,060 MW, which is the highest of its kind in the world. High head and high flow of flood discharge and dam flood vibration problems are very prominent, and the design of flood discharge and energy dissipation buildings is the best of its kind in the world.

(4) The diversion power generation system of Xiaowan Hydropower Station is located on the right bank and consists of diversion buildings, main and secondary plants, the main transformer room, tailwater buildings, and an outlet field. Six mixed-flow turbine generator sets of 700 MW are installed in the underground plant. The maximum hydraulic turbine head is 251 m, the minimum head is 164 m, and the variation is 87 m. The maximum static water pressure at the worm shell is 260 m, and the maximum dynamic water pressure is 290 m. The hydraulic turbine generator sets are characterized by the high head, high speed, large head variation, and large water thrust. The design of the underground cavern group of super large scale with stable surrounding rock and the design and manufacture of the hydraulic turbine generator set are difficult, and some of the indexes are among the top in the world.

(5) The pivot area of the Xiaowan project has steep terrain, a concentrated building layout, a narrow construction site, and high construction technical difficulty. The main project earthwork open excavation of 19.18 million m³, stone cavern excavation of 44.9 billion m³, 10.56 million m³ of concrete, and 39,800 t of metal structure manufacturing and installation. The key construction technology issues are as follows: 5 sets of 30 t, the span of 1,100 m cable engineering design and construction; the height of nearly 700 m high side slope excavation and support; dam construction intensity with a peak monthly concrete placement of 220,000 m³ and a peak annual concrete placement of 2.25 million m³; the temperature control of the construction period of the maximum pouring concrete bin surface of 26 m × 73 m of the dam. The above construction indexes ranked first in the world for similar projects.

Pan Jiazheng, an academician of the Chinese Academy of Science and Chinese Academy of Engineering, also the director of the expert committee of the Three Gorges Project and the Xiaowan Project, once said, "The hydropower dam is the king of civil engineering, and the Xiaowan Dam is the king of kings, and is deservedly the world's most complex high dam, so the construction is a reflection of modern human civilization and high-tech crystallization." As the world's first 300 m high arch dam

Xiaowan Power Plant Building

built under complex geological conditions, the Xiaowan project has challenged tradition and conventions, pioneeringly resolving the core problems of structural design, complex foundation treatment, concrete temperature control, real-time evaluation of work process, and other core problems. It established a key technology system for 300 m high arch dams, led the world's high arch dam construction to a new level, and made an important contribution to the development of the world's arch dam technology.

With a reservoir capacity of 14.9 billion m³, Xiaowan Power Station has the function of regulating water volume for many years, which not only increases the guaranteed output of downstream Manwan, Dachaoshan, and Jinghong power stations by about 1.1 million kW, which is equivalent to a new million-kilowatt peaking power station, but also makes it possible to scientifically dispatch the water volume of rivers in and outside the downstream, which helps shipping, irrigation of farmland and sand control. Since its commissioning, the project has normally been operating, and the comprehensive benefits are significant. It greatly improves local infrastructure construction and strongly promotes local economic and social development, and thus won the Second International Milestone Engineering Award for Compact Concrete Dams in 2016 and the 2019 FIDIC Engineering Project of the Year Award.

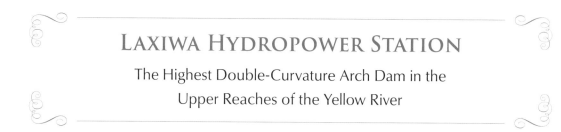

LAXIWA HYDROPOWER STATION
The Highest Double-Curvature Arch Dam in the Upper Reaches of the Yellow River

Laxiwa Hydropower Station is the second cascade of the planned large and medium-sized hydropower station in the upstream section of the Yellow River from Longyangxia to Qingtongxia, located on the main stream of the Yellow River at the junction of Guide County and Guinan County in Qinghai Province. The power station is 32.8 km away from the upstream Longyangxia Hydropower Station, 73 km away from the downstream Lijiaxia Hydropower Station, 134 km away from Xining City, Qinghai Province, and 25 km away from the downstream Guide County, with convenient external transportation.

The installed capacity of the Laxiwa Hydropower Station is 4.2 million kW. The normal storage level of the reservoir is 2,452 m, and the calibration flood level is 2,457 m. The total reservoir capacity is 1.079 billion m³; the reservoir has a daily regulating capacity of 150 million m³; the multi-year average power generation capacity is 10.223 billion kW·h, with a guaranteed output of 0.99 million kW and a rated head of 205 m.

The project pivot building consists of a concrete double-curvature arch dam (three table holes, two deep holes, and one bottom hole), a dissipation pond behind the dam, and an underground diversion

Laxiwa Hydropower Station panorama

power generation system on the right bank. The top elevation of the logarithmic spiral line double-curvature arch dam is 2,460 m, the maximum dam height is 250 m, the arch crown beam is 49 m wide at the bottom and 10 m wide at the top, which is the highest double-curvature arch dam in the upper reaches of the Yellow River. The pressure diversion pipeline is an underground buried pipe for a single machine pipe water supply, with a diameter of 9.5 m, a single pipe diversion flow of 380 m³/s, and a maximum hydrostatic pressure of 234.7 m. The underground plant cavity group is arranged in a granite body 150–466 m from the shore and buried at a depth of 230–426 m. The main and deputy plants, the main transformer switch room, and the tailwater regulator room are arranged parallel to each other in the axis. The total length of the main and deputy plants is 316.75 m, and the maximum excavation height of the main plant is 74.9 m. The tailwater system adopts the form of three machines sharing one tailwater regulating room and one tailwater cave. The impedance regulating chamber has an inner diameter of 28 m and a height of 52 m, and a tailwater gate operation room is set up between the tailwater regulating chamber and the main switch room.

The most outstanding feature of Laxiwa Power Station is its 750 kV output voltage level, which is the highest in China; the 207 m high drop metal pipeline busbar is the world's highest; the large anti-arch dissipation pond, which is the first of its kind in China; the pressure-filled water seal of the diversion cavern gate, which is the first of its kind in China and can better accommodate the installation deviation of the gate slot and ensure the smooth launching and reliable sealing of the gate; the 250 m high arch dam, which uses initial water storage for power generation, is the first of its kind in China; the shortest time used from concrete opening to water storage and power generation; the lowest investment per kilowatt in China; and the leading level of project quality, safety and civilized construction in the same industry in China.

The Laxiwa hydropower project is the largest in scale in the Yellow River basin, has the most electric power, and has good economic benefits from hydropower projects in the Yellow River Basin. It is the northern channel backbone of the power project of the "West-East Power Transmission," but also a strategic project to realize the northwest water and thermal power "bundled" and transmitted to the north China power grid. The construction of Laxiwa Hydropower Station will not only change Qinghai's resource advantages into economic advantages, but also strongly promote the economic development of Northwest China and the surrounding areas, which is of great significance to promote the overall development of society.

Xiluodu Hydropower Station

China's Second-Largest Hydropower Station

Xiluodu Hydropower Station is located in Xiluodu Gorge, which borders Leibo County of Sichuan Province and Yongshan County of Yunnan Province, 184 km from Yibin City downstream, and 770 km, 1,065 km and 1,780 km from Three Gorges, Wuhan and Shanghai respectively. The power station is mainly used for power generation, sand control, flood control, driftwood, shipping, and other comprehensive benefits. It also cooperates with the Three Gorges Project to improve the flood control capacity of the middle and lower reaches of the Yangtze River, giving full play to the comprehensive benefits of the Three Gorges Project.

The Xiluodu Double-Curvature Arch Dam project is of huge scale, with an installed capacity of 12,600 MW, second only to the Three Gorges project in the world, and tied for second in the world with the Itaipu hydropower plant. With a reservoir capacity of 11.57 billion m³, it can realize incomplete annual regulation. The power plant has a guaranteed output of 3,395 MW at the beginning of operation, and an average annual power generation capacity of 57.12 billion kW·h for many years, including 14.5 billion kW·h during the dry period. The project interception began on November 7, 2007, and the first units generated electricity in June 2013, and the construction was completed in 2015.

Xiluodu Hydropower Station panorama

The project hub is mainly composed of a barrage dam, water discharge building, water diversion, and power generation system. The barrage dam is a concrete double-curvature arch dam. The top elevation of the arch dam is 610 m, the lowest elevation of the foundation surface is 324 m, the maximum dam height is 285.5 m, the total concrete volume is about 6.7 million m³, and the total water thrust is 14 million t. The foundation is required to have a high bearing capacity, so that arch dam structure stress and stability meet the design standards. The water discharge building consists of seven table holes and eight deep holes in the dam body, four flood relief tunnels in the water pad ponds, and mountains on both sides of the dam. The maximum flood flow in the hub can reach 48,926 m³/s, of which 32,278 m³/s in the dam body and 16,648 m³/s in the flood relief cavern group. All these indexes are top in the world arch dam hubs. The diversion power system is located in the mountains on both sides of the river. The diversion system is "one machine and one pipe," the main plant and the main change room, tailwater gate well, tailwater regulating well, and other large cavity groups are compactly arranged, the tailwater system is "two machines and one hole"; the underground main plant on both sides of the river is 31.9 m wide, 75.6 m high, and 439.7 m long. It is the largest underground cavern complex in the world at present.

Xiluodu hydropower station is located in a deep mountain valley area, which is a great challenge for the design and construction technology, and many technical problems are the most difficult in China or even in the world.

(1) Construction of high arch dams in high seismic zones. The height of the concrete double-curvature arch dam is 285.5 m, which is higher than that of the Ertan arch dam (240 m) built in China and the Inguri arch dam (272 m) built abroad. The basic earthquake intensity of the dam site area is VII degrees, so the arch dam structure design and seismic safety issues are prominent.

(2) High head, large discharge, narrow river valley flood discharge energy consumption. At present, the world's largest high arch dam flood discharge power station is Ertan Hydropower Station, 39,000 MW, less than 1/2 of Xiluodu Hydropower Station.

(3) The design and construction of a huge underground cavern system. The number and size of the underground plant cavern group of Xiluodu Hydropower Station are the largest in the world. Two underground plants are symmetrically arranged on the left and right banks of the dam site, each installing nine sets of 700 MW hydro generator units. The dimensions of the main plant are 381 m × 31.9 m × 75.1 m, and the main transformer room and the large span tailwater regulator room are also arranged parallel to the plant. In addition, there are two to three flood relief tunnels on each bank of the mountain, with a single flood relief capacity of about 4,000 m³/s, forming a three-dimensional crossover with the plant system. In order to construct the diversion, three diversion tunnels of 18 m × 20 m are laid in each low elevation of the mountain on both sides of the river, and the size of its main plant, flood discharge volume, and diversion tunnel are all world-class.

(4) Long-distance high-capacity transmission technology. The transmission distance from the power station to Central China and East China is about 1,100 km and 1,800 km, respectively. It adopts DC transmission, with voltage levels of ± 500 kV and ± 600 kV, respectively, and a transmission

capacity of 12,600 MW. This long-distance, large-capacity DC transmission is the first of its kind in China and also rare in the world.

In addition, the builders also innovated and applied the "Xiluodu Arch Dam Intelligent Construction Key Technology," which pioneered the digitalization of China's 300 m high arch dam construction and realized the digital monitoring and recording of the whole process of dam construction. It plays an extremely important role in dam construction quality, progress control, and dam operation safety monitoring. Now with 6,700 sensors buried in the entire dam, relying on the "dam intelligent construction management system platform," managers can fully sense the dam's "health status," and accurately grasp the dam's "safety pulse." Zhang Chaoran, a Chinese academician, commented: "Xiluodu dam has established a set of the full process, all-round, full time, the whole life cycle of the simulation system. It can be used to calculate through simulation to master the life of the dam. Intelligent dam construction is the future of hydropower development, leading the world's high arch dam development direction." This "Xiluodu Arch Dam 300 m Intelligent Construction Key Technology" won the Second Prize in National Science and Technology Progress in 2015, which is of great significance.

Xiluodu Hydropower Station flood release

Xiluodu Hydropower Station is one of the most representative hydropower projects in the world due to its large scale and high difficulty. Xiluodu Hydropower Station, relying on the "strongest brain" of the dam, won the FIDIC Award for Engineering Project Excellence in 2016. It not only represents the highest level of intelligent construction of dams in the world, but also demonstrates the strong innovative strength of Chinese hydropower construction to the world.

GEZHOUBA HYDROPOWER STATION

The First Large Hydropower Station on the Main Stream
of the Yangtze River

Located in Yichang City, Hubei Province, 2.3 km downstream of Nanjinguan, the Three Gorges outlet of the Yangtze River, Gezhouba Hydropower Station is the first large-scale water conservancy hub built on the main stream of the Yangtze River and the shipping staircase of the Three Gorges Project. It can counter-regulate the water level downstream of the Three Gorges Hydropower Station and generate electricity by using the river fall.

Gezhouba Hydropower Station panorama

The pivot buildings from the left bank to the right bank are: left bank earth and rock dam, No. 3 ship lock, Sanjiang sand flushing gate, concrete non-overflow dam, No. 2 ship lock, concrete water retaining dam, Erjiang power station, Erjiang drainage gate, Dajiang power station, No. 1 ship lock, big river drainage sand flushing gate, right bank concrete water retaining dam and right bank earth and rock dam.

The Gezhouba Dam project was started in December 1970, and the Dajiang was intercepted in January 1981. In June of the same year, the Sanjiang locks were opened to traffic, and the first unit of the Erjiang Power Station was put into operation in July, and all the units of the Erjiang Power Station were completed in 1983. The first unit of Dajiang Power Station was connected to the grid and generated electricity in June 1986.

The dam site controls a watershed area of about 1 million km², with a total reservoir capacity of 1.58 billion m³. The concrete dam has a crest elevation of 70 m, a maximum dam height of 48 m, and a total crest length of 2,606.5 m. The Erjiang sluice gate has 27 holes, with a maximum discharge flow of 83,900 m³/s; the Sanjiang sand flushing gate has six holes, with a maximum discharge flow of 10,500 m³/s; the Dajiang sluice gate has nine holes, with a maximum discharge flow of 20,000 m³/s. The power station is of riverbed type, with two sets of 170 MW and five sets of 125 MW units installed in the Erjiang Power Station and 14 sets of 125 MW units installed in the Dajiang Power Station, with a total installed capacity of 2,715 MW and an annual power generation capacity of 15.7 billion kW·h. Navigable buildings include two waterways and three locks, among which No. 1 lock in the Dajiang waterway, and No. 2 and No. 3 locks in the Sanjiang waterway. No. 1 and No. 2 locks have effective dimensions of 280 m in length and 34 m in width, allowing large passenger and cargo ships and a 10,000 t fleet passing. No. 3 lock has effective dimensions of 120 m in length and 18 m in width, allowing 3,000 t passenger and cargo ships to pass.

The construction of the Gezhouba dam project solved some complicated technical problems. For example, the construction method of "throwing steel gabions and concrete tetrahedron to form a rock barrier to protect the bottom" has solved the problem of intercepting the flow under high flow; the use of impermeable plates and pumping measures to reduce the lifting pressure, the tooth wall to cut off the weak layer and other measures to improve the slip stability of the dam body. The dam adopts the operation method of "static water navigation, dynamic water sand flushingz' to successfully solve the problems of river planning and channel siltation. The 170 MW single-unit turbine designed and manufactured by China is one of the world's largest low-head rotary turbines in the 20th century.

XIN'AN RIVER HYDROPOWER STATION

China's First Self-Surveyed, Self-Designed, and Self-Constructed Large-Scale Hydropower Station

Xin'an River Hydropower Station is located on Xin'an River, a tributary of Qiantang River in Hangzhou City, Zhejiang Province, 170 km away from Hangzhou City. The station is mainly responsible for peak regulation, frequency regulation, and accident backup of the East China power grid, and it has comprehensive benefits such as flood control, irrigation, shipping, and breeding.

The project consists of a concrete wide-slit gravity dam, a post-dam overflow plant, and a switching station.

Xin'an River Hydropower Station was the first large hydropower station in China to survey, design, construct and manufacture its own equipment, reflecting the level of hydropower construction in China in the 1950s. In 1973, Prof. Zhao Renjun of Hohai University (formerly East China Water Conservancy Institute) and others proposed the rainfall-runoff basin model (Xin'an River model) in the course of predicting the inlet flow of the Xin'an River reservoir. In 1989, it was honored as "one of the 100 major scientific research achievements in the past 40 years since the founding of New China," and through continuous research and improvement, it has formed a complete theory of reservoir-full flow production

Xin'an River Hydropower Station panorama

flood forecasting, which has become the most widely used and the most effective basin hydrological model in China. It serves scientific research and production well and greatly enriches the teaching content. The results have been compiled into textbooks in some European and American countries, and have been applied to the National River Flood Forecasting System of the United States.

Xin'an River Reservoir is also known as "Thousand Island Lake." Together with Hangzhou and Huangshan and other tourist attractions, it has become a famous tourist destination in Shanghai and Hangzhou area. The dam site controls a watershed area of 10,480 km², with an annual runoff of 11.3 billion m³. The total reservoir capacity is 22 billion m³, with a regulating reservoir capacity of 10.27 billion m³ and multi-year regulating performance.

The barrage dam is a concrete wide slit gravity dam with a maximum height of 105 m and a total length of 465.4 m. The dam is designed for a 1,000-year return period flood and calibrated for a 10,000-year return period flood. The riverbed dam section is arranged with nine overflow table holes, and each hole is 13 m wide. The design uses the plant roof overflow method according to local conditions, and the downward flow of water is picked downstream through the plant roof via the end differential nosecone.

The total installed capacity of the power plant is 662.5 MW, with a guaranteed output of 178 MW and an average annual power generation capacity of 1.86 billion kW·h. The top of the post-dam type plant is connected to the overflow surface of the dam, with a reinforced concrete tie plate structure and the dam body, while the lower part is separated from the dam body.

More than 20,000 hm² of farmland in Jiande, Tonglu, and Fuyang cities (counties) downstream have been protected from flooding by the regulation of Xin'an River Reservoir. 11 floods larger than 10,000 m³/s have been impounded from 1960 to 1988, alleviating direct economic losses of more than 1.1 billion yuan. The reservoir upstream formed a deep-water channel, and ships can be sailed directly from the dam to She County, Anhui Province. The reservoir downstream in the dry period increases the flow, and the waterway can be improved. The reservoir has a wide water surface and abundant aquatic products. In the mid-1990s, fish production reached 35–40 million t/a, 35 times more than that before the construction of the reservoir.

YANGUOXIA HYDROPOWER STATION

The First Power Station on the Main Stream of the Yellow River

Yanguoxia Hydropower Station is located at the outlet of Salt Pot Gorge in Yongjing County, Gansu Province, 70 km away from Lanzhou City. It is the eighth cascade hydropower station in the planning of the Yellow River Longyangxia-Qingtongxia section, and the first large hydropower station on the main stream of the Yellow River, mainly for power generation and irrigation.

Yanguoxia Hydropower Station panorama

The right bank of the pivot project consists of an overflow dam, a wide slit gravity dam on the left bank, a post-dam type plant, and an irrigation diversion pipeline.

Yanguoxia hydropower station area has steep banks, narrow river channels, and rapid water. Above the dam site, it controls the watershed area of 182,800 km^2, a multi-year average flow of 899 m^3/s, annual runoff of 28.5 billion m^3, annual sand transport of 91.7 million t, the total reservoir capacity of 220 million m^3, of which the regulating reservoir capacity of 7 million m^3. It is a daily regulating reservoir.

The construction of the Yanguoxia Hydropower Station was officially started on September 27, 1958, and the first unit was put into operation on November 18, 1961, and all eight units were installed in

November 1975. Later, units 9 and 10 were expanded and installed from March 1988 to June 1990 and from February 1997 to December 1998, respectively. Now the total installed capacity of the power station is 464 MW, with a guaranteed output of 205 MW and an annual power generation capacity of 2.28 billion kW·h. It has made great contributions to the economic development and social progress of the northwest region.

In order to save concrete, the dam adopts the concrete wide slit gravity dam type. The dam axis length is 321 m, the maximum dam height is 57.2 m, and the top width is 15.9 m. The dam is designed for a 200-year return period flood and calibrated for a 1,000-year return period. The overflow section is located on the right side of the dam, with four 12 m × 10 m overflow holes, using the force pool water leap energy dissipation. The plant is located behind the dam on the left side, with a total length of 192.5 m, a width of 18.4 m, and a height of 40.3 m. It is equipped with three sets of 44 MW and six sets of 45 MW hydro generator sets, with a design head of 38 m, using a reinforced concrete worm shell, and a guaranteed output of 139 MW. Irrigation diversion pipes are located at both ends of the dam.

Yanguoxia Hydropower Station is famous for its "short construction period, low cost and high efficiency," and enjoys the reputation of "the first pearl on the Yellow River." It is also the first monument of new China to "cure the water damage of the Yellow River and develop the Yellow River water conservancy." The completion and commissioning of the power station realized the great long-cherished wish of the Chinese people to "conquer the Yellow River and benefit mankind." However, less than two years after the power station was put into operation, due to a large amount of sand in the Yellow River, coupled with the lack of effective sand drainage facilities, the reservoir capacity was drastically reduced, and the unit output could not meet the requirements. Faced with this situation, the Station people were not intimidated by the predicament. They all put themselves into the fierce labor of sand cleaning and draining with high fighting spirit, "less grass, more electricity" became the loudest slogan at that time. In the process of struggling with the Yellow River sediment, the majority of engineers and technicians gradually figured out and summed up the effective ways of safe operation of large hydropower plants on sediment-rich rivers, and the economic and technical indicators of the power station ranked in the forefront of similar power stations in China. It makes Yanguoxia Hydropower Station, thus laying a firm foundation in the world's largest sediment content on the Yellow River, setting a model for future development and construction of power stations on the Yellow River.

Liujiaxia Hydropower Station
China's First Self-Designed Large Hydropower Station with an Installed Capacity of over 1,000 MW

Liujiaxia Hydropower Station is located in Yongjing County, Gansu Province, 100 km upstream of Lanzhou City in the Liujiaxia Canyon of the Yellow River. It is the first large hydropower station with an installed capacity of more than 1,000 MW designed and built by China itself. The project is mainly for power generation and has comprehensive benefits as well, such as flood control, irrigation, ling control, water supply, and breeding.

The hub consists of three parts: water retaining buildings, flood and sand discharge buildings, and water diversion and power generation buildings.

Liujiaxia Hydropower Station panorama

The dam site controls a watershed area of 181,770 km², with an annual runoff of 27.3 billion m³ and an annual sand transport of 89.4 million t. The normal reservoir storage level is 1,735 m, with a corresponding capacity of 5.7 billion m³ and incomplete annual regulation performance.

Construction began in September 1958, and the first unit generated electricity in March 1969 and was completed in 1974. The construction of Liujiaxia Hydropower Station accumulated rich experience in the design and construction of high dams and large hydropower stations in China.

In the 1960s, Liujiaxia Hydropower Station had a number of indicators that were the first of their kind in China at that time, such as the highest concrete gravity dam in China, flood discharge buildings with flow velocities of 45–35 m/s, large underground structures, high-pressure gates, and corresponding opening and closing machines, large capacity units, 330 kV transmission and substation equipment, etc. In 1978, it was awarded the Science and Technology Achievement Award at the National Science Congress, and in 1981, it was named the National Excellent Design Project. The main dam is a concrete gravity dam with a maximum height of 147 m and a length of 204 m; the left sub-dam is a concrete gravity dam, and the right sub-dam is a loess core wall rockfill dam. The dam is designed for a 1,000-year return period flood and calibrated for a 10,000-year return period flood. Flood discharge and sand drainage buildings include three holes open spillway on the right bank, a flood relief tunnel on the right bank (converted from a diversion hole), two holes in the dam body on the left bank, and a sand drainage tunnel on the right bank. The hydropower plant is equipped with three 225 MW mixed-flow turbine units (two were converted to 250 MW in the 1990s) and one 250 MW and one 300 MW dual water intercooler unit.

In addition to the huge comprehensive benefits of flood control, power generation, and irrigation, Liujiaxia Hydropower Station can also be jointly dispatched with Longyangxia Reservoir, and the following benefits can be added to each downstream cascade. ① Increase the guaranteed output of Yanguoxia, Bapanxia, and Qingtonxia by 250 MW. During the initial operation of the Longyangxia hydropower station, the guaranteed output of Liujiaxia Hydropower Station increased from 400 MW to 550 MW, and the annual power generation capacity increased from 5.7 billion kW·h to 5.94 billion kW·h. ② Increase the flood control capacity of each downstream cascade power station: Liujiaxia Reservoir from a 5,000-year return period flood standard to the maximum possible flood (flood flow of 11,800 m³/s); Yanguoxia Reservoir from a 1,000-year return period flood standard to 10,000-year return period flood; Bapanxia Reservoir from 300-year return period flood standard to a 1,000-year return period flood.

After the joint scheduling of Longyangxia and Liujiaxia reservoirs, Liujiaxia reservoir is a counter-regulating reservoir in the upper reaches of the Yellow River, which is responsible for irrigation, flood control, and ice-jam flood prevention downstream, and can significantly reduce the flooding, and ice-jam flooding in the downstream areas of Ningxia and Inner Mongolia, and meet the needs of downstream agricultural and industrial water supply.

Spillway flood discharge

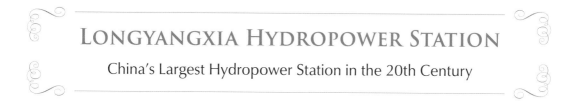

LONGYANGXIA HYDROPOWER STATION

China's Largest Hydropower Station in the 20th Century

Longyangxia Hydropower Station is located on the main stream of the Yellow River at the junction of Gonghe County and Guide County in Qinghai Province, 147 km away from Xining City, which is the uppermost cascade from Longyangxia to Qingtongxia on the upstream of the Yellow River, and is known as the "leading power station." The project is mainly for power generation, and also has comprehensive benefits such as irrigation, water supply, flood control, and ice-jam flood prevention.

Longyangxia Hydropower Station panorama

The project started in July 1978, cut off the flow in December 1979, started water storage in October 1986, the first unit generated electricity at the end of September 1987, and all units were put into operation in 1992.

The dam site controls a watershed area of 131,400 km², with an annual runoff of 20.5 billion m³ and a multi-year average sand transport of 2,490 t. The normal storage level of the reservoir is 2,600 m, with a capacity of 24.7 billion m³, making it the largest reservoir built in the 20th century in China, with multi-year regulation performance.

The project consists of a concrete gravity arch dam, left and right bank gravity piers, left and right bank concrete gravity type sub-dams, right bank spillway, plant behind the dam, etc.

The main dam is a concrete gravity arch dam with a crest elevation of 2,610 m and a maximum height of 178 m. The spillway building has two spillways on the right bank; each hole is 12 m wide and has a maximum flow rate of 4,493 m³/s; the maximum flow rates of the middle, deep and bottom holes are 2,203 m³/s, 1,340 m³/s, and 1,498 m³/s respectively. The plant behind the dam is equipped with four 320 MW mixed-flow hydraulic turbine generators with a rated head of 120 m, a guaranteed output of 589 MW, and an average annual power generation capacity of 5.942 billion kW·h for many years.

After the power station is put into operation, the comprehensive benefits are obvious. It increases the guaranteed output and power of the downstream gradient hydropower stations. For example, the four hydropower stations of Liujiaxia, Yanguoxia, Bapanxia, and Qingtongxia can increase their guaranteed output by 2.548 million kW and annual power generation by 513 million kW·h. It can improve the quality of electric power throughout the river, so that the power generation during the flood and dry periods in flat water years, dry water years, and general years is basically the same. It can increase the irrigation area of Qinghan, Gansu, Ningxia, and Inner Mongolia provinces (autonomous regions) by 315,300 hm² and improve the guaranteed irrigation rate of downstream irrigation areas. It can increase the industrial water consumption of downstream towns along the river and improve the flood control standards of important downstream industrial cities, cascade hydropower stations, and railroad trunk lines.

Longyangxia Hydropower Station represents the high level of domestic hydropower engineering construction in the 1980s. At that time, it was not only famous at home and abroad for the highest dam (178 m), the largest reservoir capacity (24.7 billion m³), and the largest generator unit capacity (320 MW), but also for its remarkable social benefits, economic benefits and "leading" position in the management of the Yellow River and the unique advantages of developing the parent power station of the upper reaches of the Yellow River hydropower resources have attracted the world's attention.

XIANGHONGDIAN HYDROPOWER STATION
China's First Self-Designed Concrete Gravity Arch Dam

Xianghongdian hydropower station is located in the upper reaches of the Xipi River, a tributary of the Huai River, at the junction of Yu'an, Huoshan, and Jinzhai counties in Liuan City, 44 km from Liuan City and 120 km from Hefei City. It is a large hydropower project mainly for flood control and irrigation, combined with power generation, shipping, aquaculture, tourism, etc.

The dam construction started in April 1956 and was completed in July 1958. All buildings were completed, and power was generated in 1961. The dam has a concrete volume of about 280,000 m³. The dam site controls a watershed area of 1,400 km², with an average multi-year runoff of 1.02 billion m³ and a total reservoir capacity of 2.632 billion m³, ranking among the top five reservoirs in the western part of Anhui Province. The designed irrigation area is 333,000 hm².

Xianghongdian Hydropower Station panorama

The pivotal buildings include a concrete gravity arch dam, flood relief tunnel, power generation tunnel, power plant, and cableway facilities for lifting bamboo and timber over the dam.

The dam is the first gravity arch dam designed and constructed by the People's Republic of China after its establishment. It is a fixed-center, single-curvature concrete gravity arch dam, 87.5 m high, 361 m long at the top of the dam, 6 m thick at the top, and 39 m thick at the bottom. The maximum compressive stress was 2.94 MPa, and the maximum tensile stress was 0.735 MPa. In 1983, the seismic analysis was performed using the three-dimensional finite element method, and the maximum tensile stress was 1.57 MPa.

There is one flood relief tunnel located on the right bank, which is arranged on the downstream side of the water diversion and power generation cave. The power station is located on the right bank and is connected by one main tunnel to four branch tunnels to divert water for power generation. The plant is arranged on the ground level with four sets of 10 MW units installed. The first unit started to generate electricity on September 15, 1959, and all four units were put into operation on April 6, 1961.

In order to improve the comprehensive power generation, flood storage, and drought resistance capacity of Xianghongdian Hydropower Station, the construction of the hybrid pumped storage power plant in Xianghongdian was started on December 16, 1994, and was completed in June 2000. It is the first pumped storage power station in Anhui Province, jointly invested by Anhui Electric Power Company and Anhui Energy Investment Company. With its installed capacity of 2 × 40 MW, it forms a hybrid pumped storage power station with a total installed capacity of 120 MW together with Xianghongdian Hydropower Station, providing 228 MW of peaking and valley filling capacity for the Anhui power grid.

GEHEYAN HYDROPOWER STATION
China's First Upper-Gravity-and-Lower-Arch Composite Dam

Geheyan Hydropower Station is located in Changyang County, Hubei Province, on the main stream of Qingjiang River, a tributary of the Yangtze River, 207 km from Enshi City and 50 km from Gaobazhou Hydropower Station. The project is mainly for power generation and comprehensive benefits such as flood control and shipping.

Geheyan Hydropower Station panorama

The project was started in early 1987, and the first unit generated electricity in June 1993, all four units were connected to the grid in November 1994, and it was basically completed by the end of 1994. The hub project passed the completion inspection and has operated well.

The dam site controls a watershed area of 14,430 km², accounting for 85% of the total watershed area, with an average multi-year flow of 403 m³/s, an annual runoff of 12.6 billion m³, and an average multi-year sand transport of 9.71 million t. The designed flood standard for the main hydraulic buildings of the hub is a 1,000-year return period, with a flood peak flow of 22,800 m³/s and a corresponding reservoir level of 203.14 m. The calibrated flood standard is a 10,000-year return period, the peak flood flow is 27,800 m³/s, and the corresponding reservoir level is 204.59 m. The normal storage level of the reservoir is

200 m, the corresponding capacity is 3.12 billion m³, the dead water level is 160 m, and the capacity of the reservoir is 1.975 billion m³. total reservoir capacity 3.44 billion m³, reservoir capacity of 0.5 billion m³ reserved for flood control. It can not only reduce the flood peak in the lower reaches of Qingjiang River, but also stagger the encounter with the flood peak of the Yangtze River, reduce the chance of using the flood diversion project of Jingjiang River and postpone the time of flood diversion.

The elevation of the top of the hill on both sides of the dam site is about 500 m. The width of the river is 110–120 m during the dry period. The lower 50–60 m bank slope of the river valley is steep, while the upper part of the river valley is steep on the right and gentle on the left, which is an asymmetric canyon. The base stratum of the dam is Cambrian Shilongdong Tuff, with dense and hard lithology, and the bedrock is more developed with faults, interlayer shear zones, and karst systems. The basic intensity of the earthquake in the dam area is VI degrees, and the design intensity is VII degrees.

The installed capacity of the hydropower plant is 1,200 MW, with a guaranteed output of 187 MW and a multi-year average annual power generation of 3.04 billion kW·h. It also undertakes the task of frequency regulation and peak regulation of the central China power grid.

The pivot project consists of a concrete gravity arch dam, a water discharge building, a right bank diversion type hydropower station, and a left bank vertical boat lift. The top elevation of the dam is 206 m, the total length of the dam is 665.45 m, and the maximum height of the dam is 151 m. Gravity dam sections are arranged on both banks, and gravity piers are set up on the left shoulder of the dam at an elevation of 120–138 m. The riverbed is a three-centered single-curved, top-heavy, and bottom-arch composite gravity arch dam with an outer arc radius of 312 m and a downstream slope of 1:05–1:0.7. The top of the arch in the middle of the riverbed is 181 m in elevation, gradually decreasing to both banks, 150 m from the left bank to the top of the gravity pier and 160 m from the right bank to the shore, with grouting of the horizontal joints below the top of the arch to form an arch dam of different grouting heights, and no grouting of the horizontal joints above the top of the arch, which is the working state of the gravity dam, with a downstream slope of 1:0.7 and a transition section between the two. For the weak structural surface that affects the stability of the arch seat on both sides of the river, it is treated with measures such as slip-resisting keys, force-transmitting columns, and strengthening the drainage of the mountain.

The spillway structure is centrally located in the middle of the riverbed of the dam, with a length of 188 m. There are seven table holes, four deep holes, and two bottom holes which are also used for diverting the flow. The table hole weir top elevation is 181.8 m, and the size of the orifice is 12 m × 18.2 m. The elevation of the bottom of the deep hole is 134 m, and the size of the orifice is 4.5 m × 6.5 m. The bottom elevation of the bottom hole is 95 m, and the size of the orifice is 4.5 m × 6.5 m. All kinds of orifices are controlled and operated by arc gates, and flat maintenance gates are installed upstream of them. The surface hole body type adopts an asymmetric wide tail pier, and the deep hole body type adopts a narrow slit pick flow nose can. The flood discharge capacity of the table hole is 17,050 m³/s and 19,000 m³/s under the design and calibration conditions, respectively, and the maximum discharge capacity of the hub is 24,000 m³/s. The seepage control curtain line is 1.5 km long, with a total feed of 251,700 m.

The hydropower plant is located on the right bank and is a diversion-type ground plant with four 9.5 m diameter tunnels connected to 8 m diameter pressure steel pipes, with a single tunnel connected

Aerial view of Geheyan Hydropower Station

to four 300 MW mixed-flow turbine units, respectively. The tunnels are lined with prestressed concrete. The excavation of the plant and pressure steel pipe forms a high slope of 170 m, which is treated by partial replacement of concrete, installation of prestressing anchor beam, and strengthening of mountain drainage.

The navigable building is located on the left bank and is China's first 300 t class high lift vertical ship lifter over the dam. The designed maximum annual cargo capacity is 3.4 million t, with a total lift of 124 m. The project is divided into two stages: the first stage is part of the leading edge of the dam water retaining, with a lift of 42 m; the second stage is located on the left bank downstream river bank, with a lift of 82 m, which is connected with the middle staggered shipping channel and the downstream river channel. The ship hoist adopts a full balance wire rope winch system; the effective water size of the ship-bearing compartment is 42 m × 10.2 m × 1.7 m, with total water weight of 1,400 t.

The completion of Geheyan Hydropower Station has made an important contribution to the development of Qingjiang River hydropower energy, reducing the burden of flood control on the Yangtze River, improving water transportation in the mountainous areas of southwest Hubei Province and the development of minority areas in southwest Hubei Province.

FUZILING HYDROPOWER STATION

The First Hydropower Station in the Huai River Basin

Fuziling Hydropower Station is located in the upper reaches of the Huai River tributary of the Pihe River. It is in Huoshan County, Dabieshan Mountains, in western Anhui Province, 17 km away from Huoshan Town. The dam is located 2.5 km south of Fuziling Town, hence the name. It is one of the first large water conservancy projects built to cure the flooding of the Huai River and is known as the first hydropower station in the Huai River basin. Fuziling Hydropower Station was built in January 1952 and completed in November 1954. The hydropower station is mainly for flood control, taking into account the comprehensive benefits of irrigation, power generation, shipping, and fishery.

Fuziling Hydropower Station was the first backbone project of the Huai River control at the beginning of the establishment of the People's Republic of China. It was designed by China itself as a large continuous arch dam with an advanced international level at that time. It is known as "the first dam in New China," "the first dam in Asia," and "the tallest dam in the Far East," and has a milestone significance in the history of water conservancy and hydropower construction in the People's Republic of China.

The dam site controls a watershed area of 1,270 km², with an average multi-year flow of 49.5 m³/s. The hub is designed for a 100-year return period flood and calibrated for a 1,000-year return period flood.

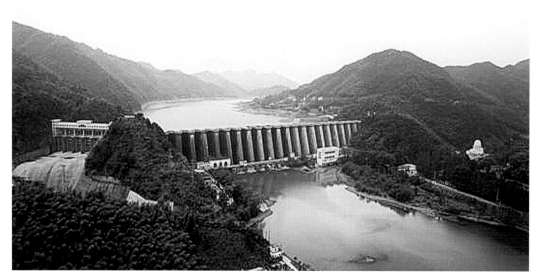

Fuziling Hydropower Station panorama

The design flood level is 129.44 m, the calibration flood level is 130.84 m, the total reservoir capacity is 496 million m³, the flood limit water level is 117.56 m, and the beneficial water level is 124.96 m, the corresponding reservoir capacity is 384 million m³. The hub project includes four parts: a dam, spillway, water transmission pipe, and power station.

The dam is a reinforced concrete continuous arch dam, 74.4 m high (75.9 m after raising), 1.8 m wide at the top, and 510 m long, divided into three sections: the east dam, the continuous arch dam, and the west dam. The east dam is a gravity dam, 52 m long; the west dam is a gravity dam at the lower part and a flat dam at the upper part, 45 m long; the central section of the continuous arch dam is 413.5 m long, consisting of 20 stacks and 21 arches. The top elevation of the dam is 128.46 m, which was raised to 129.96 m in 1983 (excluding the 1.1 m wave wall). The open-top spillway is located at the shoulder of the right bank dam, the crest elevation of the weir is 112.56 m, the top width is 63.6 m, with six holes, single hole width of 10.6 m. Each hole is installed with double-capacity roller plate steel gates, the maximum flood discharge capacity is 7,500 m³/s, and the 100-year return period flood flow is 5,000 m³/s.

The dam is equipped with nine water transmission pipes, three of which are installed in the No. 13, No. 14, and No. 15 stacks for flooding and irrigation, with a maximum design flow of 225 m³/s. The remaining six are water diversion pipes for power generation. Irrigated arable land area is 200,000 hm² (including Muzitan Reservoir). Fuziling Reservoir, together with Xianghongdian and Meishan reservoirs, can irrigate nearly 670,000 hm² of farmland. The power station has seven installed units, installed in the new and old plants, with a total installed capacity of 31 MW and a multi-year average annual power generation of 124 million kW·h.

Fuziling Reservoir with Arch Dam

The river valley bottom is about 200 m wide, and the bedrock is mostly quartzite and granite, with hard lithology, no major faults, but more developed fissures. It stood the test of earthquake and flooding during and after construction.

As an important part of the Huai River management project, the Fuziling Reservoir impounded a large amount of flood water from the Pihe River, which not only greatly improved the flood control standard of the middle and lower reaches of the Pihe River, but also played a role in assisting the main stream of the Huai River to store flood water. The Fuziling Dam was the first reinforced concrete arch dam designed and constructed by China itself, and the whole design process was carried out in the context of exploration and construction, with no experience nor norms to follow. Wang Huzhen and other water specialists led a group of young and middle-aged technical members to learn and do at the same time. They overcome the transverse earthquake stress analysis, dam stabilization and grouting of the dam base, and other major technical problems in their design. The completion of the Fuziling Dam has amazed the water conservancy engineering community at home and abroad, and has won a great reputation for the New China, turning a new page in the cause of large-scale water management in China. Mr. Toran, the chairman of the International Committee on Dams, called the Fuziling Dam "an international first-class earthquake-proof continuous arch dam." The director of Leningrad Hydroelectric Design Institute praised: "The continuous arch dam is good. Chinese engineers are great!"

Fuziling Reservoir is deeply engraved with the figures and footprints of the first generation of the Chinese Communist Party and State leaders. When Chairman Mao Zedong visited Anhui, he went to the provincial museum to see the model of Fuziling Reservoir; Committee Chairman Zhu De and Marshal Liu Bo Cheng visited the reservoir in person and gave great care to the construction and development of Fuziling Reservoir. Many firsts have been created here: the first dam in New China, the first reinforced concrete support pier dam in New China, the first 100-meter high dam in Anhui Province, the first hydropower station in Huai River Basin, the first hydro generator set in New China, and the first 110 kV transmission line in Anhui Province. There is a rich historical and cultural heritage here: Guo Moruo, Liu Haisu, Guan Shanyue, Wu Zuoren, Chen Dengke, and other literary figures have left many works for Fuziling, "Fuziling University," which has trained a large number of water conservancy and hydropower talents for China, and the spirit of the old generation who managed the Huai River. In the long volume of history, Fuziling Reservoir has written the glory and splendor that belongs not only to itself but also to the People's Republic of China. Over the past 60 years since the reservoir was built, it has intercepted more than 200 floods of all sizes, created huge social and economic benefits, and played an important role in the economic and social development of the basin and the improvement of people's livelihood.

PUDING HYDROPOWER STATION
China's First Compact Concrete Arch Dam

Puding Hydropower Station is located in Puding County, Guizhou Province, south of Wujiang River, 125 km away from Guiyang City, mainly for power generation, but also for water supply, irrigation, breeding and tourism, and other comprehensive benefits. The project was intercepted on December 15, 1989, the concrete was poured in December 1991, the dam was completed on May 30, 1993, and the power was connected to the grid in June 1994.

Puding Hydropower Station panorama

The dam site controls a watershed area of 5,871 km², with an annual runoff of 3.88 billion m³. The river valley is asymmetric U-shaped, and the dam is built on a base surface of tuff. The basic intensity of the earthquake in the project area is less than VI degrees, and the design is based on VI degrees of protection. The normal storage level of the reservoir is 1,145 m, with a total capacity of 421 million m³. The design flood standard of the main hydraulic buildings of the hub is a 100-year return period, and the calibration flood standard is a 500-year return period.

The project consists of a compact concrete arch dam, an open spillway at the top of the dam, a sand flushing hole, and a river bank type power plant. The dam is a fixed-center, variable-radius, variable-center-angle, equal-thickness hyperbolic asymmetric arch dam with a crest elevation of 1,150 m, a dam height of 75 m, and a total crest length of 195.67 m. The total amount of concrete in the dam body is 137,000 m³, of which 103,000 m³ is compact concrete, accounting for 75.2% of the total. The hydropower station is equipped with three hydroelectric generating units, with a total installed capacity of 75 MW and an annual power generation capacity of 340 million kW·h.

The Puding Arch Dam is the first compacted concrete arch dam constructed by applying full-section rolling technology in China. The main features of the project are as follows.

(1) The dam is constructed without construction joints, using integral, thin-layer, through-bin, full-section crushing, and filling, revolutionizing the traditional construction process of splitting joints, blocks, and columns for normal concrete arch dams.

(2) Induced joints are provided in appropriate parts of the dam body, and a repetitive grouting system is provided in the joints so that in case the temperature stress exceeds the tensile strength of the dam concrete, it is first pulled away from the induced joint area for grouting treatment.

(3) The dam body is impermeable using the compact concrete itself. In order to improve the overall impermeability of the dam body compact concrete and prevent horizontal seepage along the compact level, special treatment measures were taken at each compact level to increase the level bond strength so that the dam body compact concrete has a seepage resistance of W10 or more.

(4) By mixing high fly ash and reducing the amount of cement, the dam concrete temperature stresses were effectively controlled, and the crack resistance of the dam body was improved.

The Puding Arch Dam is the only test dam of the 100-meter compact concrete arch dam in China. The overall quality is much higher than that of the compact concrete arch dams built during the same period, and in many mechanical indicators and physical properties, it is as good as or even better than that of the normal concrete dams built during the same period.

HUNAN TOWN HYDROPOWER STATION

China's Highest Buttress Dam

Hunan Town Hydropower Station is located on the Wuxi River, a tributary of the Qiantang River in Quzhou City, Zhejiang Province, about 25 km down from Huangtankou Hydropower Station. It is one of the cascade power stations for the development of hydropower resources in the Wuxi River, with comprehensive benefits of power generation, flood control, irrigation, shipping, and water supply.

Hunan Town Hydropower Station panorama

Hunan Town Hydropower Station construction was started in 1958, paused in 1961, and resumed in 1970. The first unit generated electricity in 1979, and the station was completed in 1980. The expansion project was started in October 1994, and the unit was put into trial operation on November 1, 1996, and then into normal operation.

The dam site controls a watershed area of 2,197 km² with an average multi-year flow of 83.4 m³/s. The normal reservoir storage level is 230 m, with a corresponding storage capacity of 1.582 billion m³ and a total storage capacity of 2.03 billion m³, having the capacity of incomplete multi-year regulation. Through reservoir regulation and joint operation with Huangtankou Hydropower Station, the dry period output of downstream Huangtankou Hydropower Station can be increased.

The pivot building includes the barrage dam, diversion tunnel, diversion-type ground plant, plant behind the dam, and switching station. The barrage is a concrete trapezoidal pier dam with a maximum height of 129 m, a total length of 440 m, an upstream slope of 1:0.2 and a downstream slope of 1:0.68, a top width of 7 m. The thickness of the head of the dam is gradually thinned from the bottom to top, with a 5 m wide stiffening wall between the two piers.

The flood relief building is located in the middle of the riverbed, including five overflow table holes at the top of the dam and four bottom holes in the support pier of the overflow dam section, with a maximum discharge flow of 11,000 m³/s. The total installed capacity of the power station is 270 MW, of which the old power plant on the right bank, built in 1980, has an installed capacity of 170 MW. In 1996, a 100 MW unit was expanded using a 5.4 m diameter steel pipe reserved in the left bank dam section. The power station has a guaranteed output of 52.1 MW and an annual power generation capacity of 540 million kW·h.

The construction of Hunan Town Hydropower Station was the first in China to adopt a concrete trapezoidal dam; the first to adopt a crescent-shaped inner reinforced rib steel fork pipe; the first to adopt SF composite materials in the gate design, and successfully developed SF3A type supporting slide on the inlet gate and SF2C type supporting hinge sleeve on the spillway arc gate; and new technologies and structures such as ring beam vertical column type machine pier were adopted in the plant design. In 1999 Hunan Town Hydropower Station expansion project, a compact, reasonable layout of buildings and the use of a fully enclosed plant structure successfully solved the problem of the dam flood fogging on power plant operation; the use of low-temperature reservoir water spray after the air supply ventilation design; successfully achieved the goal of "unmanned, less manned" and remote computer monitoring operation. The operation of the unit is controlled by advanced technology such as AGC (Automatic Generation Control Operation).

MEISHAN HYDROPOWER STATION
China's Highest Concrete Continuous Arch Dam

Meishan Hydropower Station is located in the upper reaches of Shihe River, a tributary of Huai River, at the junction of Hubei, Henan, and Anhui provinces. It is in Jinzhai County, Anhui Province, 130 km north of the mouth of the Shihe River into the Huai River. It is a large water conservancy and hydropower project with comprehensive benefits such as flood control and irrigation, as well as power generation. It is also a key project self-designed and self-constructed by China during the first five-year plan.

Meishan Hydropower Station panorama

Construction work began in March 1954, the main dam project was basically completed in January 1956, and the reservoir began to store water in 1958.

Meishan Reservoir is a multi-year regulating reservoir with a flood control capacity of 1.065 billion m³ and a designed irrigation area of 255,300 million hm². The dam is a reinforced concrete continuous arch structure with a maximum height of 88.24 m. It is the highest concrete continuous arch dam in China, consisting of 15 stacks and 16 arches. The left and right ends of the continuous arch dam are connected to a section of gravity dam and a hollow gravity dam. The total length of the dam axis is 443.5 m. The installed capacity of the power station is 4 × 10 MW.

The pivot building includes a reinforced concrete continuous arch dam, right bank open spillway, right bank flood relief tunnel, No. 9 arch spillway bottom hole, and behind dam type power plant.

In the early stage of reservoir storage, on September 28, 1962, water storage started and continued to completion. The highest water level is 125.56 m. The high water level continued until the early morning of November 6, the right bank of the dam shoulder bedrock fissure suddenly leaked a lot of water. The total amount of measured water leakage reaches 70 L/s. An unblocked consolidation grout hole at the base of No. 14 stack sprays water outward. The measured pressure head reaches 31 m, equivalent to the reservoir water level on 82% of the head at this location. On the right bank of the dam roof and arch, stack cracks also appeared in a number of places, including No. 15 arch crown inside the cracks from the top of the arch extended down to 28 m long, the maximum seam width of 6.6 mm. After emptying the reservoir and inspection, up to more than 100 m long, about 30–80 m wide range of bedrock local sliding and tension crack was found near the front edge of the bedrock contact surface on the right bank of the arch platform between No. 13 and No. 16 stack. After that, a series of engineering measures such as consolidation and curtain grouting, additional gravity pier of the right bank stack, support wall, bedrock prestressing anchorage, and drainage facilities behind the dam were taken to repair and reinforce the dam. It withstood the test of two major floods in 1969 and 1991, and 40 years of operation. The first periodic safety inspection of the Meishan multiple arch dam was conducted in 1991–1992 and confirmed compliance with current specifications and was rated as normal.

Over the past 60 years since its completion, Meishan Hydropower Station has brought into play tremendous comprehensive engineering benefits. In terms of flood control, it has taken on the important task of storing for the main stream of Huai River, especially overcoming the floods in 1969, 1991, and 2003, ensuring the safety of people's lives and properties in the downstream towns, and making outstanding contributions to ensuring the safe flooding of Huai River and reducing the middle reaches of Huai River to enable flood storage areas; in terms of irrigation, as the main source of water for the Pishihang irrigation area, it irrigates 255,000 hm^2 of farmland in 5 counties in the downstream of Anhui and Henan Provinces, providing a stable and safe harvest in drought and flood; in terms of power generation, it is an important power source for the power grid in western Anhui, playing an irreplaceable role in the development of the Anhui power grid and ensuring the safety of the power grid. All kinds of benefits, especially flood control, irrigation, and power generation, play a very important role in the economic development and social stability of the western part of Anhui Province and even the whole province.

LUBUGE HYDROPOWER STATION
China's First Project for International Public Bidding

Located at the junction of Luoping County, Yunnan Province, and Xingyi City, Guizhou Province, on the Huangni River, a tributary of the Nanpan River in the Pearl River system, Lubuge Hydropower Station is a mixed development hydropower station, the last cascade power station on the Huangni River, a tributary of the Nanpan River on the left bank of the upper reaches of the Pearl River. The power station is developed singularly for power generation. The power station was a key project during the "Sixth Five-Year Plan" and "Seventh Five-Year Plan" of China, and was the first power station built by China in the early 1980s with World Bank loans and international bidding, introducing advanced foreign equipment and technology. It is known as the "window" power station of China's hydropower infrastructure opening to the world.

Lubuge Hydropower Station panorama

The project started in November 1982, cut off the flow in 1985, the first unit generated electricity in December 1988, and passed the completion inspection in December 1992.

The dam site controls a watershed area of 7,300 km² with an annual runoff of 5.17 billion m³. Both the reservoir area and the dam area are canyon river sections with steep slopes and rapid flow, with bedrock outcrops of Permian, Triassic, and Carboniferous tuffs and sand shales. The basic intensity of the earthquake in the dam area is VI degrees, and the dam was designed according to the VII degree fortification. The normal storage level of the reservoir is 1,130 m, with a corresponding capacity of 111 million m³, and the dead water level is 1,105 m, with a regulating capacity of 75 million m³ with seasonal regulation performance. The power station is developed in a hybrid way, with a dam at the first part to congest the water level to form a head of 85 m; a long tunnel is excavated to divert water, with a concentrated drop of 287 m and a total head of 372 m. The installed capacity is 600 MW, with a guaranteed output of 85 MW and a multi-year average annual power generation of 2.845 billion kW·h.

The project consists of three parts: the head of the pivot, the water diversion system, and the plant area. The head of the pivot includes the barrage dam, water discharge building, and sand drainage tunnel. The barrage is a heart wall rockfill dam with a maximum height of 103.8 m and a crest elevation of 1,138 m. The heart wall material is taken from the nearby mixture of fully weathered sand shale and residual soil, with a top width of 5 m and a bottom width of 38.25 m. The downstream slope ratio on the heart wall is 1:0.175, and the upstream and downstream sides are respectively provided with one 4 m thick layer and two 5 m thick back filter layers. The slope ratio of the dam upstream and downstream is 1:1.8, with a total filling volume of 2.22 million m³. The drainage building is a two-hole open spillway on the left bank, with a maximum discharge flow of 6,424 m³/s; the left bank flood tunnel has a maximum discharge of 1,995 m³/s; the right bank flood tunnel has a maximum discharge flow of 1,658 m³/s. There is one sand drainage tunnel with a design discharge flow of 300 m³/s.

The water diversion system is located on the left bank, including the water inlet, pressure diversion tunnel, pressure regulating well, and pressure pipeline. The riverbank inlet is located 500 m upstream of the dam, with a floor elevation of 1,091 m. The diversion tunnel is 8 m in diameter and 9,387 m long, with a maximum diversion flow of 230 m³/s. The pressure-regulating well is a different type with the upper chamber, followed by two buried inclined wells and high-pressure mains, connected to four branch pipes. The plant is located at the exit of the canyon, and the plant is underground, with four 150 MW single-capacity mixed-flow turbine generators, with a maximum head of 372.5 m and a rated head of 312 m. The main transformer room and switch station are underground.

During the construction, a series of new technologies and techniques were adopted, such as crushing weathered material core wall by vibration bulges, excavation of vertical step steep slope and high spillway slope by bolting, excavation of diversion tunnel by drilling and blasting, smooth blasting, full section pouring of needle beam steel mold and rock wall crane beam for the concrete lining. During the construction period, the Lubuge project set 14 national records and won the "National Excellent Survey (Gold) Award," 'National Excellent Design (Gold) Award," and "Luban Award for Construction Engineering." It was awarded as one of the 100 classic and high-quality projects on the 60th anniversary of the founding of the People's Republic of China.

The construction of the Lubuge Hydropower Station can be described as tortuous. As early as the 1950s, the relevant state departments began to arrange for the survey of Huangnihe River, Kunming Hydropower Survey and Design Institute undertook the design of the project, and the former Ministry of Hydropower embarked on the construction of Lubuge Power Station in 1977, and the 14th Bureau of Hydropower began to build roads and prepare for the construction. However, at the early stage of project construction, the project faced serious challenges, specifically: firstly, due to a serious lack of funds, the project had not been able to start, and the progress of the preparatory works was slow; secondly, there was a lack of competitiveness. Since 1949, China's large-scale hydropower project construction had been in a self-management system, with no modern concept of project management; thirdly, the management level was low. The hydropower construction, like other industries, only copied the 1950s Soviet industrial enterprise management model, and this led to low management level, inflexible operation mechanism, backward management mode, low level of technology application, and low technology content; fourth, backward technology. Single management mode resulted in backward technology and low construction efficiency; fifth, personnel team quality was not good, and the enterprise was overloaded. These challenges restricted the development of the Lubuge power Station construction. Limited by the traditional organizational mode, the project construction was stalled for a time, and the preparatory works progressed slowly, so the project was delayed for seven years. In 1984, the former Ministry of Hydropower introduced the World Bank loan, which immediately solved the fund-lacking problem. The introduction of the World Bank loan also brought new ideas of fund management and project management. The international competitive bidding method effectively reduced the project cost and improved the efficiency of capital utilization, which led to the project implementation pattern of "one project, two systems, and three party construction."

The introduction of a competition mechanism in the Lubuge project and the advanced and efficient construction practice had a significant impact on the management system, labor productivity, and compensation distribution of China's engineering construction at that time. It promoted the reform of China's hydropower construction management system, which was called "Lubuge Shockwave." The "Lubuge Shockwave" brought about the liberation of ideas, and Chinese hydropower construction took the lead in implementing owner responsibility, bidding for contracting and construction supervision system, and popularizing project construction experience. The new hydropower construction system was gradually established, the self-run system of the planned economy was put to an end, and the results of the reform gradually appeared. This new management model brought about a great improvement in efficiency, accelerated the process of hydropower development in China, and promoted the reform of China's hydropower construction management system. The "Lubuge Shockwave" attracted widespread attention and had far-reaching effects. After this, national construction projects of all sizes began to try out the bidding system and contract management. Its influence has long gone beyond the hydropower system itself and has had a strong impact on people's thinking. It is an important milestone in the history of China's hydropower construction reform and occupies a place in the history of China's reform and opening-up.

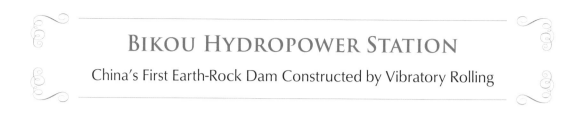

BIKOU HYDROPOWER STATION

China's First Earth-Rock Dam Constructed by Vibratory Rolling

Bikou Hydropower Station is located on the Bailong River, a tributary of the Jialing River in the Yangtze River system in Wenxian County, Gansu Province, China. The project is mainly used for power generation, mostly supplying power to the northern Sichuan and southern Shaanxi power systems and the Bikou area in Gansu Province. It also has comprehensive benefits such as flood control, irrigation, shipping, breeding, and tourism.

Bikou Hydropower Station panorama

The main project started in 1969, cut off the flow in March 1971, and impounded water at the end of 1975, the first unit generated electricity in 1976, and the remaining two units were put into operation one after another the next year.

The dam site controls a watershed area of 26,000 km² with an annual runoff of 9.06 billion m³. The normal reservoir storage level is 704 m, with a total reservoir capacity of 521 million m³ and a regulating reservoir capacity of 221 million m³, with seasonal regulation performance. It is designed to irrigate 590 hm² of downstream farmland, with an annual over-wood of 500,000 m³. The installed capacity of the power station is 300 MW, with a guaranteed output of 78 MW and a multi-year average annual power generation of 1.463 billion kW·h.

The project consists of a loam core dam, right bank spillway, flood tunnel, timber passage, left bank flood tunnel, sand discharge tunnel, power generation cavern and plant, etc. It is a second-class project.

Bikou loamy core wall earth-rock dam is designed for a 500-year return period flood, calibrated for a 5,000-year return period, and protected for a 10,000-year return period flood. The dam's top elevation is 710 m, the maximum dam height is 101.8 m, and the dam's length is 297.36 m. The dam shell is made of compact rock, stone slag, gravel, and other materials. The total filling volume is 3.966 million m³. In order to prevent the dam from being overtopped by unstable waves, an L-shaped concrete wave retaining wall of 5.3 m in height is installed at the top of the dam. The deepest depth of the sand and pebble cover of the dam base is 34 m, and two concrete impermeable walls have been set under the core wall, of which the latter one is as deep as the bedrock, plus the total depth of 68.5 m reaching into the core wall.

Hub water release buildings are right bank open spillway, maximum discharge 2,310 m³/s; right bank pressureless flood relief tunnel, 13 m wide, 15 m high, maximum discharge 2,250 m³/s; left bank pressurized flood tunnel, diameter 10.5 m, maximum discharge 1,711 m³/s; left bank pressurized sand drainage and flood tunnel, diameter 4.4 m, maximum discharge 296 m³/s. The flow velocity of each drainage building is above 34 m/s, which is a high-speed water flow.

The water diversion and power generation system is located on the left bank, including the water inlet, diversion tunnel, and pressure regulating well. The diameter of the diversion tunnel is 10.5 m, and the maximum diversion flow is 480 m³/s. The regulating well is an impedance-type square shaft. The main plant is 83 m long, 20 m wide, and 46.82 m high, and is equipped with three 100 MW mixed-flow turbine generators with 4.1 m diameter and 73 m rated head, and three longitudinal timber transfer machines at the right dam head with 900 m³ per shift, which are no longer in use.

Bikou Dam was the first earth-rock dam higher than 100 m constructed in China. It is the first time in China's construction to adopt vibratory rolling technology, to use anchor spraying support for the rock surrounding the weak underground cavern, and to apply the "dragon head" way to convert the diversion cave into a flood relief tunnel.

YILI RIVER CASCADE HYDROPOWER STATION

China's Earliest High-Head Cascade Hydropower Station

Yili River Cascade Hydropower Station is composed of four hydropower stations: Maojiacun, Shuicaozi, Yanshuigou, and Xiaojiang. It is a cross-basin development cascade hydropower station built in the 1950s in Huize County, Yunnan Province. Yili River cascade hydropower station mainly generates electricity, while Maojiacun reservoir also has irrigation utility.

Yili River is a tributary of Jinsha River, with a basin area of 2,558 km², a total length of 122 km, a natural water surface drop of 2,000 m, and an average annual rainfall of 900 mm. The average multi-year flow rate at the Shuicaozi dam site is 19.3 m³/s. The downstream section of the Yili River below Shuicaozi is nearly parallel to the flow direction of the Jinsha River and is close to it. But the riverbed elevation of the Yili River is about 1,380 m higher than that of the Jinsha River. So the inter-basin water diversion development method is adopted at Shuicaozi station. The total installed capacity of the Yili River cascade hydropower station is 321.5 MW, with an average annual power generation capacity of 1.6 billion kW·h.

Maojiacun hydropower station and reservoir are built on the main stream of Yili River, which is a full-stage regulating reservoir; the barrage of Shuicaozi Hydropower Station is built on the main stream of Yili

Maojia Village Earth Dam

River, and the underground plant for power generation is built in the mountain outside the main stream of Yili River, and the tailwater for power generation is injected into the regulating pool of Yanshuigou hydropower station outside the watershed of Yili River. Yanshuigou Hydropower station uses the tailwater of Shuicaozi Hydropower station for power generation. Xiaojiang Hydropower Station, using the tailwater of Yanshuigou hydropower station, incorporates the Xiaojiang River as well to increase the water used for power generation. The tailwater, after power generation, is directly discharged into the Jinsha River.

The construction of the Shuicaozi Hydropower Station started in 1956, and the rest of them were all completed in 1972, one after another. Yanshuigou and Xiaojiang Hydropower Stations are developed for high head, with the maximum head reaching about 629 m, and installing bucket-type hydro generators with a single capacity of 36 MW. In the Yili River Cascade Hydropower Station Project, the barrage of Maojiacun Hydropower Station is a clay core wall earth dam with a maximum height of 82.5 m, while the barrage of Shuicaozi Hydropower Station is an overflow concrete gravity dam. The power plants of the four cascade stations are all underground. The main characteristics of each power station are shown in the following table.

The Main Characteristics of the Yili River Cascade Hydropower Station

Hydropower Station	Maojiacun	Shuicaozi	Yanshuigou	Xiaojiang
Controlled Watershed (km²)	868	1,233	1,233	1,386
Multi-year Average Flow (m³/s)	15.9	19.3	19.3	21.8
Maximum Head (m)	77	79	629	628.2
Installed Capacity (MW)	16	17.5	14	144
Annual power Generation Capacity (Hundred Million kW·h)	0.73	0.92	7.16	7.19

With the Yili River Cascade Hydropower Station serving peak regulation of Central Yunnan power grid and supporting the accident backup, it has irrigated 4,900 hm² of farmland and protected 900 hm² of farmland from flooding.

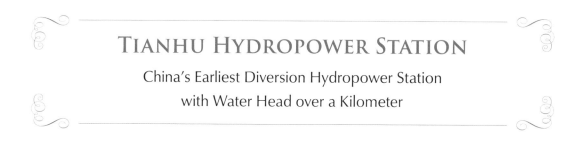

TIANHU HYDROPOWER STATION

China's Earliest Diversion Hydropower Station
with Water Head over a Kilometer

Tianhu Hydropower Station is located in the upper reaches of the Yima River in the Xiangjiang River system, 35 km from Quanzhou County, Guilin City, Guangxi Zhuang Autonomous Region.

Construction of the first phase began in July 1989 and was completed in April 1992; The second phase began in December 1994 and was completed in July 1999.

Tianhu Hydropower Station panorama

The project consists of four systems: water storage (diversion), water transmission, power generation, and power transmission. The highland water storage (diversion) system, with the Tianhu Reservoir and the Haiyangping Reservoir at the core, consists of 13 small and medium-sized reservoirs forming an interlinked, up-and-down reservoir group and water diversion and transmission network. The reservoir group is located in highland areas 1,400 m above sea level, with a total controlled catchment area of 43.67 km² and a total reservoir capacity of 34.24 million m³.

The water transmission system consists of two parts: the part before the front pond is a channel and unpressurized tunnel, which brings water from each reservoir to the front pond; after the front pond, the water is diverted to the power generation system by unlined pressure shafts and pressure steel pipes.

The power station is designed to generate a hydrostatic head of 1,074 m. The pressure waterway system consists of a 1,428 m elevation of Dawangshan tunnel and Erwangshan tunnel, 1,400 m elevation of T-shaped tunnel, vertical shaft and inclined shaft, 827 m elevation of the flat cave, inside open pipes and outside open pipes connected, with a total length of 4,500 m. The power station is designed to have an installed capacity of 60 MW, and is built in two phases, each with an installed capacity of 30 MW. The multi-year average power generation capacity is 185 million kW·h.

By the end of the 20th century, Tianhu hydropower station was the highest head diversion hydropower station in China and even Asia, and its design water pressure and PD value exceeded the scope of China's design specifications. Under the special circumstance where the specifications were only for reference at that time, in order to ensure the reliability of the pressure steel pipe, the fuzzy optimization theory is used for the first time to establish the mathematical model for optimal design and for the membrane stress area of the steel pipe and fork pipe, increasing the allowable stress by 3% than those designed in a conventional way in the case of super high head and super specification index. And the pressure steel pipe design was optimized, so the design is both safe and economical. In the operation of the power station, the pressure steel pipe has withstood the full dumping load test without abnormalities. Its economic and social benefits are obvious. In the structure design, new technology was applied: pressure steel tube separated dynamic support and super high head dissipation well drainage energy dissipation. Good static and dynamic performance sealing material of PTFE asbestos packing and special di-aluminum were firstly used respectively for expansion joint and manhole. New construction craft of thick steel plate, small diameter roll round without leaving the head roll solved the problem of forming difficulty of pressure steel pipe with high crank and thick steel. Meanwhile, submerged automatic arc welding is adopted, and the plate is preheated by far infrared before welding and treated with the electric furnace at constant temperature after welding to eliminate welding slag, bubbles, residual stress, and cold bending stress, which ensures the quality of steel pipe production and installation.

LIULANGDONG HYDROPOWER STATION

China's First Groundwater Runoff Hydropower Station

Located in Xindian Township, Qubei County, on the border of Honghe and Wenshan Autonomous Prefecture in Yunnan Province, Liulangdong Hydropower Station is one of the 16 hydropower stations invested in and built during China's "second five-year plan," and the first hydropower station in China to generate electricity directly from groundwater in karst areas. Construction of the power station began in February 1958, and it was put into operation in February 1960 and completed in March of the same year as an underground hydroelectric power station with an installed capacity of 12,500 kW × 2.

Liulangdong Hydropower Plant

The Liulangdong River is an underground river, which flows from the karst cave and converges into the Nanpan River after 5.2 km of open flow, with a drop of 104 m and an average slope drop of 2%, with a river basin area of 846 km², of which the open flow section has a basin area of 39 km². The power station draws the underground water of Liulangdong, a right bank tributary of the Nanpan River, to generate electricity. The surface water converges with the underground water in the natural karst cave— Liulangdong, so the underground water abundance is rare in the world. The cavern reservoir is formed by intercepting and plugging the leakage, and the 3,368.33 m long underground water diversion tunnel leads the water to the main plant for power generation, with the tailwater discharged into the Nanpan River.

The dam site controls a watershed area of 807 km² with an average multi-year flow of 22.6 m³/s. The normal storage level of the reservoir is 1,086 m, with a total storage capacity of 271,000 m³ and an effective storage capacity of 237,400 m³, which is a daily regulated groundwater reservoir. The installed capacity of the power station is 25 MW.

The pivot head of the power station (including the spillway, sand flushing sluice, cave plugging works, water intake, and underground reservoir) is arranged at the entrance of the Liulangdong, using dry blockwork to seal the cave and concrete and reinforced concrete slabs to prevent seepage, raising the water level in the cave and forming a daily regulated underground water reservoir. The former limestone cave outlet and the lower cave entrance are used to construct sand-flushing gates and spillways, which constitute buildings for water congestion, flood discharge, and sand flushing.

The spillway is arranged on the limestone rock base at the lower cave entrance of the former Liulangdong, with a gravity overflow weir, the top elevation of which is 1,083 m, and the height of the dam (gate) is 11.6 m. An arc-shaped gate is set at the top of the weir, and the original design orifice size (width × height) is 6 m × 3 m. In 1970, the orifice size was reduced to 4 m × 1.5 m due to the instructions from the higher level to raise the water level and overload the operation. After the first round of inspection, according to the expert group's opinion, it was restored to the original design orifice size of 6 m × 3 m. The downstream adopts flip flow energy dissipation.

The sand flushing gate is arranged on the right side of the spillway, with a reinforced concrete structure. The bottom plate elevation of the gate is 1,068 m, and the height of the gate body is 10.5 m. The bottom hole is used to release the flow equipped with an arc-shaped working gate, with an 1.8 m × 1.8 m (width × height) orifice.

The power station inlet is located upstream of the upper cave drop, with a floor elevation of 1,068 m. The inlet is tunneled to take water directly from the cave.

As of 2011, the Liulangdong Hydropower Plant had generated a total of 8.279 billion kW·h of electricity, but the plant's equipment had aged badly, and there were many safety hazards. In order to meet the province's demand for electricity for economic and social development, improve the reliability and automation level of power plant operation, ensure long-term safe and stable operation, and at the same time make fuller use of the rainy season hydro energy resources, it was decided to add two new 15,000 kW hydro generator units next to the old power plant to replace the two 12,500 kW units in service. On November 9, 2010, the construction project of the new Liulangdong Hydropower Plant officially started and was completed in 2013, with the two old units shutting down.

BAISHAN HYDROPOWER STATION

The Largest Hydropower Station in the Northeast of China

Baishan Hydropower Station is located in Huadian City, Jilin Province, on the upper main stream of the Songhua River, 39 km downstream from Hongshi Hydropower Station and 250 km from Fengman Hydropower Station. It is the largest hydropower station in the northeast region, in the middle of the north-south network of the northeast power system, and is mainly responsible for system peak regulation, frequency regulation, and accident backup tasks. A combined operation with the downstream Fengman Hydropower Station will increase the latter's guaranteed output by 19 MW and annual power generation capacity by 46 million kW·h.

Baishan Hydropower Station panorama

The project was started in 1958, suspended in 1961, and prepared for resumption in 1971 in two phases, with three units of the first phase of the project generating electricity in 1984 and two units of the second phase of the project going into operation in 1992.

The dam site controls a watershed area of 19,000 km², with an annual runoff of 7.4 billion m³ and an annual sand transport of 1.05 million t. The reservoir has a normal storage level of 413 m, a total storage capacity of 676.6 billion m³, and performs multi-year regulation. Through reservoir dispatching, when there is a 100-year return period flood, the downstream flow of Fengman Reservoir can be reduced by 500–1,000 m³/s, and thus the downstream flood disaster is mitigated.

Baishan Dam is a three-centered circular gravity arch dam with a maximum height of 149.5 m and a crest arc length of 676.5 m. The dam is designed for a 500-year return period flood and calibrated for a 500-year return period flood, with the maximum possible flood protection. All flood water is released by the dam body, and the flood relief facility consists of four overflow table holes and three middle holes arranged at intervals. The maximum flow rate of the overflow meter hole is 8,880 m³/s, and the maximum flow rate of the middle hole is 4,110 m³/s. The meter hole and middle hole both adopt the nose-can pick flow energy dissipation, and each pick can adopt a different pick angle and plane diffusion angle to form a three-dimensional comprehensive dispersion energy dissipation with mutual penetration of the pick flow tongue, horizontal diffusion, and vertical stratification to meet the downstream limited scouring and siltation requirements. Baishan Hydropower Station has a total installed capacity of 1,500 MW, implemented in two phases, with 900 MW installed in phase 1 and 600 MW installed in phase 2, with a guaranteed output of 167 MW and a multi-year average annual power generation capacity of 2.037 billion kW·h. The right bank phase 1 underground plant is located 90 m downstream of the dam in the mountain, 121.5 m long, 25 m wide, and 54.25 m high, with three mixed-flow turbine generator sets, a single capacity of 300 MW, and a rated head of 112 m. The main transformer and switching station are arranged underground, and the regulating wells are cylindrical. The main transformer and switching station are located underground, and the regulator well is of cylindrical type. The second phase of the left bank is a ground-level plant with two turbine generators of the same type as those on the right bank. The ground-level switching station is located on the left side of the plant. The maximum head of the power station is 126 m, the design head is 110 m, and the minimum head is 86 m.

Baishan Hydropower Station is the first cascade in the development of the west-flowing Songhua River, jointly dispatched with the downstream group of Hongshi Hydropower station and Fengman Hydropower Stations to improve the efficiency of full-range operation. It has played a prominent role in the operation of the northeast power system in terms of peaking and frequency regulation and accident backup, and has been of great benefit since its commissioning.

WUJIANGDU HYDROPOWER STATION
The Highest Gravity Arch Dam in China's Karst Region

Located in the middle reaches of the Wujiang River, a tributary of the Yangtze River in Zunyi City, Guizhou Province, Wujiangdu Hydropower Station is the first large-scale project to develop the hydropower resources of the Wujiang River and is also the main power station in Guizhou's power grid. In addition to supplying electricity for industry and agriculture in Guizhou Province, it also sends electricity to Chongqing and acts as a peaking agent for the grid.

Wujiangdu Hydropower Station panorama

Construction preparations began in 1970. The first unit started its power generation in 1979 (during which work was suspended for two years), and the project was completed in 1983. The dam site controls a watershed area of 27,800 km², with an annual runoff of 15.8 billion m³. The normal storage level of the reservoir is 760 m, with a total storage capacity of 2.3 billion m³. The dam is designed for a 500-year return period flood and calibrated for a 5,000-year return period flood.

The project consists of a concrete gravity arch dam, dam spillway, powerhouse and switching station, left and right bank flood relief tunnels, left and right side of the dam sand drainage and flood relief middle hole, right bank emptying tunnel, and boat lifts.

The maximum height of Wujiangdu Gravity Arch Dam is 165 m, with a crest elevation of 765 m and a crest arc length of 395 m. There are six holes of the spillway on the dam face, the middle four holes adopt the plant top pick flow method, and the left and right holes are ski-way type pick flow methods. The inlet of the left and right bank spillway is located at No. 16 and No. 3 dam sections, respectively, with a 9 m wide and 10 m high orifice and an arc-shaped working door.

The main plant of the power station is located behind the dam, with a fully enclosed structure and three hydro generator sets of 210 MW single capacity, with a guaranteed output of 202 MW and an average annual power generation capacity of 3.34 billion kW·h. The sub-plant is located in the lower cavity of the overflow surface on the upstream side of the main plant.

The Wujiangdu Gravity Arch Dam is the first high dam built in China in a karst area. The dam site is located in the Triassic Yulong Mountain limestone formation with karst development, and a short distance downstream of the dam is the nine-stage beach shale formation. The dam base is protected by a high-pressure cement grout curtain, which is connected to the upstream sand shale water barrier to solve the reservoir leakage problem. The total length of the seepage curtain line is 1,020 m. The deepest reaches 200 m below the riverbed, with a total curtain area of 189,000 m² and a maximum grouting pressure of 6 MPa, with 4 and 5 grouting tunnels on the left and right, respectively. Due to the narrow river valley and large flood flow, the spillway, main plant, sub-plant, and switching station overlapped on the elevation. Four spillway table holes, two middle holes, one ski-way spillway on each side, and one spillway hole on each bank are used for joint flood discharge, and the outlet is used for energy dissipation. Four table holes are used to pick the flow in front of the plant, and the flood flow is picked downstream over the plant behind the dam. The other holes and caverns are laid at different elevations and at different distances below the dam, so that the fall point of each pick flow is pulled apart longitudinally and dispersed when the flood is released to achieve the purpose of dispersing energy dissipation and reducing scouring, which has proven to be effective over the years. In order to avoid cavitation of the flow channel structure caused by high-velocity water flow, gas doping facilities are used in the flow channel. The concrete project adopts an artificial aggregate processing system with an annual production capacity of 2 million tons, which solves the problem of a lack of local natural sand and gravel.

Fengtan Hydropower Station is located downstream of Youshui, a tributary of the Yuanjiang River in the Yangtze River system in Yuanling County, Hunan Province, and is a dam-type hydropower station, which was put into operation in 1978. The project is mainly for power generation, and also has comprehensive benefits of flood control, shipping, irrigation, etc. It is the first hollow-bellied gravity arch dam in China.

Fengtan Hydropower Station panorama

The watershed area above the dam site is 17,500 km², with an annual runoff of 15.9 billion m³. The normal storage level of the reservoir is 205 m, with a total storage capacity of 1.733 billion m³ and a regulating storage capacity of 0.33 billion m³, which has seasonal regulating performance. Through the regulation of the reservoir, it can reduce the flood disaster downstream, irrigate 4,400 hm² of farmland downstream, improve the shipping conditions upstream and downstream of the reservoir, and also provide favorable conditions for the development of farming in the reservoir area.

The project consists of a concrete barrage, a spillway building, a plant, and a rafting channel.

In order to adapt to the specific situation of the narrow river valley, high flood flow, and complex geological conditions of the dam site, the project adopted a special arrangement of concrete hollow-belly gravity arch dam, overflow at the top of the dam, and a plant in the hollow. The maximum height of the Fengtan hollow-belly gravity arch dam is 110 m, which is the highest hollow-belly dam in the world at that time. The total length of the hollow is 255.8 m, the maximum hollow height is 40.1 m, the width is 28.7 m, and the volume of the whole hollow is 184,000 m³. The hollow on the left side of the arch crown has the main plant and installation yard; in the right hollow, there is a 220 kV switching station.

The spillway building consists of 13 spillway table holes and one emptying spillway bottom hole. The spillway dam section has an arc length of 255 m and a centripetal angle of about 61°. The net width of each spillway hole is 14 m, and the energy is dissipated by using interlocking high and low nosecans to pick up the flow, including seven low nosecans and six high nosecans. These high and low cans pick flow air collision can eliminate the downstream flow of 50% of the energy and reduce the scour depth of the downstream riverbed. Hydraulic model tests and more than ten years of operation practice prove that the energy dissipation and anti-scouring effect are good. The right side of the overflow dam section is equipped with one bottom hole for flood discharge, with a maximum discharge capacity of 1,236 m³/s.

The power station plant is equipped with four 100 MW mixed-flow hydro generators with a rated head of 73 m, a guaranteed output of 103 MW, and a multi-year average annual power generation capacity of 2.043 billion kW·h.

The dam crossing raftway uses a vertical plus ramp lift, allowing 50 t class vessels to pass.

ERTAN HYDROPOWER STATION
China's Highest Double-Curvature Arch Dam in the 20th Century

Ertan Hydropower Station is located at the junction of Yanbian and Miyi counties in Panzhihua City, the southwest border of Sichuan Province, downstream of Yalong River. It is built mainly for power generation, but also for other comprehensive purposes. The dam site is 33 km away from the confluence of the Yalong River and Jinsha River and 46 km away from Panzhihua City. It is the first hydropower station of Yalong River Hydropower Base, the upstream is Guandi Hydropower Station, and the downstream is Tongzilin Hydropower Station. The normal storage level of the reservoir is 1,200 m, with a total storage capacity of 5.8 billion m³, a multi-year runoff of 52.7 billion m³, and a regulating storage capacity of 3.37 billion m³.

Ertan Hydropower Station panorama

Ertan Hydropower Station was included in the national plan in 1987 and was built by the central and local joint venture, partly financed by a World Bank loan. Construction officially started in September 1991, with the first unit generating electricity in July 1998. It was completed in 2000.

Ertan flood discharge photo

The power station hub consists of a barrage dam, flood relief buildings, energy dissipation buildings, a power plant, and other components. The barrage is a concrete parabolic double-curvature arch dam, consisting of 39 sections, with a maximum height of 240 m and an arc length of 774.69 m, which is the first 200 m high arch dam built in China. The spillway is located on the right bank of the Yalong River. The spillway is a shallow, short-inlet, dragon-head open-flow spillway, consisting of an inlet section, a dragon-head section, a straight section of the cavern, and an outlet picket section, with a design flow of 7,400 m³/s and a calibration flow of 7,600 m³/s. The energy dissipation building consists of a water pad pond and two dams. The total installed capacity of the power station is 3.3 million kW, with a guaranteed output of 1 million kW and a multi-year average power generation capacity of 17 billion kW·h. It is the largest power station built and put into operation in China in the 20th century.

In the engineering survey and design work, a lot of scientific research work has been carried out for the key technical problems of the project, including "calculation methods and design principles of high concrete dams," "Evaluation of the stability of the rock body of high concrete dams and research on the quality of the available rock body," "High concrete dam flood discharge energy dissipation research," and other topics of national key scientific and technological research. During the construction process, advanced international technology was adopted in the impermeable cofferdam wall, foundation treatment process, dam concrete system, underground structure excavation and support, and the manufacture of a large hydro generator set. For example, two 4 × 4.5 m³ concrete mixing plants manufactured by CIFA from Italy, each with a production capacity of 360 m³/h, were used in the construction; three 30 t radial cable machines were used for concrete placement; in the construction of large caverns and high side walls of the underground project, a large number of 175 t prestressed anchor cables and sprayed steel fiber concrete spray were used in consideration of high ground stress and possible rock burst.

The Ertan project ranks top in many aspects in China and the world. ① First in China: China's first high dam over 200 m (dam height of 240 m; realizing the leap from 150 m to 240 m, taking the first steps in China's high dam construction); China's largest underground plant cavern group (also the largest in Asia); the largest power station built and put into operation in the 20th century (total installed capacity of the power station is 3,300 MW); the largest single turbine generator unit capacity in China: 550 MW, a big leap from 335 MW to 550 MW; the first domestic hydropower project to fully implement international competitive bidding; the largest flood relief cave, section height 13.5–14.9 m, width 13 m, the maximum flow speed of 45 m/s, all ranking first in the country; water inlet height 80 m, the height of the pressure chamber 70 m, all ranking first in the country. ② The world's first: the World Bank has the largest loan for a single project; the dam bears a total load of 9.8 million t and the power plant design discharge capacity of 22,480 m³/s, top in the world's high dams; the diversion cavern cross-sectional area (23 m high, 17.5 m wide) is the world's largest. Ertan Hydropower Station is hailed by its foreign counterparts as the most outstanding water conservancy project completed and put into operation in China in the 20th century. It marks a new historical stage in the development of China's hydropower industry, and its brilliant achievements and inspiration are a great treasure for the Chinese nation. It has enabled Sichuan and Chongqing to say goodbye to years of electricity tension and laid the foundation for economic development in the 21st century.

Ertan Hydropower Station won the 10th National Gold Award for Excellent Engineering Design, the 8th National Gold Award for Excellent Engineering Survey and the 1st National Environmental Friendly Engineering Award, and the 6th China Civil Engineering Zhan Tianyou Award. The successful completion of the Ertan Hydropower Station has brought light to the start of the construction of Xiaowan, Jinping-I, and other domestic extra-high arch dams, forming the "Ertan model" that can be referred to as the design and hub layout of domestic thin arch dams are almost all copies or adaptations of Ertan model. Ertan project also has trained a large number of talents, and thus made a good technical reserve for the design of Xiluodu, Jinping, Pubugou, Dagangshan, and other mega power stations later.

SHAPAI HYDROPOWER STATION

The World's Highest Compacted Concrete Arch Dam

Shapai Hydropower Station is located in Wenchuan County, Sichuan Province, on the Caopo River, a first-class tributary of the Min River, about 136 km from Chengdu City. Construction of the project started on June 16, 1997, and was fully completed in December 2003.

The dam site is located at the northwestern edge of the Sichuan Basin, with a narrow river valley in a roughly symmetrical V-shape, and concave slopes and strip ridges on the left and right banks of the downstream. The rock integrity of the dam base is good, and the riverbed cover is 30–40 m thick. The basic seismic intensity is VII degree. The normal storage level is 1,866 m, and the dead water level is 1,825 m. The total reservoir capacity is 18 million m³, with seasonal regulation performance.

Shapai Compact Concrete Arch Dam

The main buildings of the hub are the compact concrete arch dam, the flood relief cavern, the diversion tunnel, and the power plant house. The dam is a three-centered round single-curved arch dam with a height of 130 m, which is the world's highest compacted concrete arch dam under construction at the end of the 20th century. The dam does not have a flood vent, and two flood tunnels are set up on the right bank. The total installed capacity is 36 MW, with a multi-year average annual power generation capacity of 178 million kW·h.

The Shapai arch dam, with a compact concrete volume of 383,000 m³, is a key science and technology research project in China from 1990 to 2000, and has accumulated design, scientific research, and construction experience for the construction of a 100-meter-high compact concrete arch dam. The main features of the project are as follows.

(1) A three-centered circular single-curved arch dam was selected to facilitate shoulder stability and bias the direction of arch thrust toward the interior of the mountain. The height of the padded seat is 12.5 m, the arc length of the dam top centerline is 250.25 m, and the maximum center angle is 92.48°.

(2) In order to facilitate construction, the dam does not have a flood relief orifice. Only two flood relief tunnels are set on the right bank, and one is a vortex shaft flood relief tunnel converted from a diversion cave, design head of 88 m, a design flow of 244 m³/s. The internal energy dissipation rate is above 73%.

(3) The dam adopts a structure of 2 induced joints combined with two conventional transverse joints, and buries high-density polyethylene cooling water pipes to pass water for cooling in the parts of the dam poured during the high-temperature season; the repeated grouting technique is used to allow multiple repeated grouting of the open induced joints.

(4) The dam is impermeable by itself using secondary-matched C20 compact concrete rich in cementitious material on the waterward side, with a net slurry laid between the layers. A 2 mm thick polymer coating sheathing was installed below the normal water storage level as a secondary impermeability.

The Shapai Dam withstood the test of landslide in the 5·12 Wenchuan earthquake. It was safe and unharmed, becoming the "strongest" dam in the earthquake. The dam has normally been operating since then, providing strong energy support for the resumption of production.

NANSHUI HYDROPOWER STATION

China's Best Rockfill Dam with Directional Blasting Technology

Nanshui Hydropower Station is located in Nanshui, Shaoguan, Guangdong Province, 16 km from the county town of Ruyuan Yao Autonomous County, the second largest artificial lake in Guangdong Province and the hydropower station with the largest installed capacity in Shaoguan City.

The project underwent a major demolition on December 25, 1960, and was impounded for power generation in 1969.

Nanshui Hydropower Station panorama

The dam site controls a watershed area of 608 km², with an average multi-year flow of 33.4 m³/s, a design flood flow of 4,190 m³/s, a total reservoir capacity of 1.218 billion m³, and an installed capacity of 75 MW. The mountains on both sides of the dam site are high, the river valley is V-shaped, the water surface is about 15–30 m wide, and the geological conditions are quartz sandstone.

The dam is constructed by directional blasting and piling, with a height of 81.8 m, a crest length of 215 m, a crest width of 8 m, and a bottom width of 430 m. The slope of both the upper and lower rock piles is 1:3, with a design blasting volume of 1.14 million m³ and an average piling height of 65 m.

The construction adopts the blasting plan of laying a large pack mainly on the right bank, combined with a small auxiliary pack, with a total charge of 1,394 t. After testing, the throwing volume is 1.05 million m³, throwing the effective amount of 1 million m³ on the dam; the center of the rock pile body is only a few meters away from the designed, and the blasting orientation is accurate; the average height of the rock pile is 62.3 m, the bottom width is 420 m, the upper and lower rock piles slope is 1:3, which basically matches with the design, and the blasting and piling are well performed; the impact of blasting on the bedrock damage is small, and it is controlled within the envisaged damage range; the average void rate of the blasted rock pile is less than 30%, and the compactness is high. After the reservoir was built and operated normally, it was tested by three floods and proved that the quality of the dam was good. It becomes the best project for large-scale directional blasting dam construction in China.

The water quality of Nanshui Reservoir is crystal clear and blue, with green hills stretching along the banks, and the bamboo and wooden buildings of Yaozhai Village hidden in the greenery. In the northwest of the reservoir, on a peninsula with an area of about 667 hm², the first hunting ground in southern China was built. Take a boat from the reservoir pier to the hunting ground, and you can enjoy the scenery of the reservoir area on the way.

View of Nanshui Reservoir I

View of Nanshui Reservoir II

View of Nanshui Reservoir III

View of Nanshui Reservoir IV

TIANSHENGQIAO-I HYDROPOWER STATION

The World's Largest Dam in Panel Area and Rockfill Volume in the 20th Century

Tianshengqiao-I Hydropower Station is located on the main stream of Nanpanjiang River, and is the leading power station of the Hongshuihe River cascade development. The right bank of the dam site is Longlin County, Guangxi Zhuang Autonomous Region, and the left bank is Anlong County, Guizhou Province. The power station is 240 km away from Guiyang City, its upstream is about 62 km from Lubuge Hydropower Station on the tributary of Nanpanjiang River, and its downstream is about 7 km from Tianshengqiao-II Hydropower Station. The construction of the power station started in June 1991, the flow was cut off at the end of 1994, the first unit generated electricity at the end of 1998, and the project was completed in 2000.

Tianshengqiao-I Concrete Panel Rockfill Dam panorama

The dam site controls a watershed area of 50,139 km² with an average multi-year runoff of 19.3 billion m³. The total reservoir capacity is 10.257 billion m³, which is an incomplete multi-year balancing reservoir. The main buildings are designed for the once-in-a-1,000-year maximum possible flood calibration.

The hub consists of a concrete panel rockfill dam, an open bank spillway, an emptying tunnel, a water diversion system and ground plant, and other major buildings. Tianshengqiao-I Hydropower Station is a concrete panel rockfill dam with a maximum height of 178 m, ranking second in the world for built panel dams at the end of the 20th century; the length of the top of the dam, the volume of the dam filling and the panel area all rank first in the world. The top elevation of the dam is 791 m, the top length of the dam is 1,104 m, and the filling volume of the dam is about 18 million m³, of which 14 million m³ of filling material comes from the excavated slag material of the spillway.

The spillway placed in the right bank pass has the characteristics of large scale, large discharge volume, and high velocity, consisting of a diversion channel, overflow weir, drainage tank, pick flow nosecan, and shore protection works. The spillway weir leading edge width is 81 m, with five holes 13 m × 20 m arc gate. By the end of the 20th century, it was China's largest in scale and the amount of discharge of the shore spillway, with a maximum flow rate of 45 m/s, with gas doping corrosion reduction facilities in the relief tank.

The emptying tunnel is located on the right bank and has various functions such as water diversion, bypass, emptying, etc. The maximum discharge volume is 1,766 m³/s. It is controlled by an arc gate with a total water pressure of 87,350 kN, which is the largest hydraulic retractable water stop in China by then.

The total installed capacity of the hydropower station is 1,200 MW, with a guaranteed output of 403.6 MW and an average multi-year power generation capacity of 5.25 billion kW·h when the four machines are operated jointly.

The plant is located on the left bank downstream of the dam and uses a single machine and single pipe to divert water. The water diversion system is arranged on the left bank and consists of a diversion channel, intake tower, tunnel, and pressure steel pipe. The overlying rock of the tunnel section is thin and a fault passage zone, so the new technology of the post-tensioning method with bonded prestressed concrete lining is adopted. The main transformer is arranged upstream of the plant.

Since its operation, the Tianshengqiao-I Hydropower Station has played an important role in promoting economic development and social progress in Yunnan, Guizhou, Guangdong, and Guangxi provinces.

SHUIBUYA HYDROPOWER STATION
The World's Tallest Concrete Panel Rockfill Dam

Shuibuya project is located in the middle reaches of the Qingjiang River, a tributary of the Yangtze River in Badong County, Enshi, Hubei Province, and is the leading project of the main stream of Qingjiang River. Upstream 117 km from Enshi City, downstream 92 km from Diheyan Hydropower Station, this project is mainly used for power generation, flood control, shipping as well as working with other water hubs.

Effect picture of Shuibuya Hub Project Concrete Panel Rockfill Dam on Qingjiang River

Its construction started in January 2002, the river was intercepted in 2002, the diversion hole was gated in October 2006, the emptying hole was gated in April 2007, the generating unit operated in July 2007, and ended in August 2008 with all four units connected to the grid. The normal storage level of the reservoir is 400 m, with a total capacity of 4.58 billion m³. Four mixed-flow turbine units are installed, with a single capacity of 400 MW, totaling a capacity of 1,600 MW that guarantees 310 MW output to generate an average yearly power capacity reaching 3.92 billion kW·h.

The project is a large water conservancy hub. The main buildings are a concrete panel rockfill dam, riverbank spillway, right bank underground power plant, and emptying hole. The concrete panel rockfill dam is the highest panel dam in the world, with a maximum height of 233 m, a crest elevation of 409 m, an axial length of 660 m, and an upstream and downstream slope of 1:1.4. The riverbank spillway is arranged on the left bank and consists of a diversion channel, a control section, a drainage channel section (including a pick nose) and a downstream anti-wash section. The downstream anti-flushing section adopts the structure of an anti-poaching wall. The venting hole is arranged on the right bank for reservoir emptying, middle and late diversion, and water supply downstream during the construction period. The pressurized cave is 530.24 m long, with a diameter of 9.0–11.0 m. The unpressurized cave is 532.63 m long, with a clear space of 7.2 m × 12.0 m, and is in the shape of a city-gate cave. The diversion tunnel of the underground power station adopts one machine in one hole, with an average length of 387.9 m and an inner diameter of 6.9–8.5 m in a circular section; the size of the underground plant is 168.5 m × 23 m × 67 m (length × width × height); the tailwater hole also adopts one machine in one hole, with an average length of 313.18 m and an inner diameter of 11.3 m in a circular section.

The topography on both sides of the pivot is steep, and the environmental and geological conditions in the section near the dam at the head of the reservoir and in the dam area are relatively complex. The environmental and geological conditions in the dam site area are poor, with more dangerous rock bodies distributed around. Therefore, a number of studies have been carried out on landslide management, soft rock formation, energy dissipation forms, etc., and good results have been achieved.

On December 26, 2018, Shuibuya Hydropower Station passed the national completion inspection and acceptance, becoming the first conventional large-scale hydropower station to be accepted under the new national regulations on project completion and acceptance. Especially in response to the 2016 Qingjiang River Basin mega-flood and the 2017 mega-autumn flood, the Shuibuya Hub Project accomplished significant flood control benefits. Shuibuya project marks the closure of the construction of the Qingjiang River Basin cascade power station and the completion of the Qingjiang River Basin development and construction mission. The project has been in normal operation and management, and its benefits are being given full play.

WUDONGDE HYDROPOWER STATION

The World's Largest Single-Capacity Hydroelectric Generating Unit

Wudongde Hydropower Station is a large hydropower station in southwest China, located on the main stream of Jinsha River at the junction of Huidong County, Sichuan Province, and Luquan County, Yunnan Province, and is a very large water conservancy project in the lower reaches of Jinsha River, mainly for power generation, combined with functions in flood control, sand control and improvement of downstream shipping conditions, etc. It is the fourth-largest hydropower station in China in terms of installed capacity and the seventh-largest megawatt-class hydropower station in the world. It is also a national key project for the implementation of the West-to-East Power Transmission Project. The Wudongde Dam is a concrete double-curvature arch dam, the world's first extra-high arch dam with the application of low-heat cement concrete cast for the entire dam, and the world's thinnest 300 m class

Wudongde Hydropower Station panorama

extra-high arch dam with a height of 270 m, a bottom thickness of 51 m, an average thickness of 40 m, and a thickness-to-height ratio of only 0.19.

Wudongde Hydropower Station is the first stage of the four cascade power stations (Wudongde, Baihetan, Xiluodu, and Xiangjiaba) planned for the development of the lower reaches of the Jinsha River, and is the third in scale in the Jinsha River Hydropower Base, smaller than Xiluodu Hydropower Station and Baihetan Hydropower Station. The normal storage level is 975 m, the total reservoir capacity is 7.408 billion m³, the regulating reservoir capacity is 3 billion m³, the flood control reservoir capacity is 2.44 billion m³, the total installed capacity is 10.2 million kW, the single unit is 850,000 kW, with six units on the left and right bank. It is the world's largest hydro generator unit that has been put into operation. The average annual power generation capacity is 38.91 billion kW·h, and the static investment in the project is 33.6 billion yuan.

Wudongde Hydropower Station is one of the important projects in the West-East Power Transmission Project, meeting the demands for electricity for the economic development of east and central China. In terms of flood discharge, the Wudongde Hydropower Station has made use of the deep-water pad and the better slope conditions of the hard rock lateral valley, and has adopted a combined flood discharge scheme of five table holes, six middle holes, and three flood relief caverns on the bank of the dam, which has achieved good results. In addition to seismic and flood resistance, the main challenge of the Wudongde Dam comes from the interior of the dam—the anti-cracking joints—because of its "slim body," which, however, is required to carry the water storage of a mega hydropower plant of 10 million. The thinner the dam, the harder it can resist cracking. In order to prevent cracking, the dam body needs to be continuously raised, and the interval between the pouring of each unit should not exceed 14 days. However, the dam site is a dry and hot river valley with high ambient temperature, and the temperature difference between the inside and outside is very likely to lead to concrete cracks. In order to effectively control temperature, the project adopted the intelligent water ventilation 2.0 system to cool the concrete of the dam, creating a new mode of real-time temperature control, automatic control, and precise control, which perfectly realized the goal of pouring a seamless dam.

Wudongde Hydropower Station is a milestone project in China's progress towards becoming a country with strong hydropower, and the Jinsha River Basin Cascade Development Model fully reflects the Three Gorges Group's independent innovation, a national technology, and Chinese solutions. The Wudongde Hydropower Station Project is undoubtedly another great national project built by tens of thousands of builders and scientific workers through more than 1,000 days and nights of hard work, and is the best practice of General Secretary Xi Jinping's request to "always climb new peaks in science and technology."

Wudongde Hydropower Station Unit 3

Many world firsts were created during the construction of the Wudongde Hydropower Station.

- The world's first ± 800 kV UHV flexible DC transmission project.
- The world's largest single-station capacity flexible DC transmission project (5,000 MW).
- The world's first UHV flexible DC transmission project using a hybrid full-bridge and half-bridge valve set.
- The world's first UHV flexible DC transmission project with high-end valve sets and low-end valve sets connected in series.
- The world's first long-distance, high-capacity flexible DC transmission project with a transmission distance of over 1,000 km.
- The world's first flexible DC transmission project with overhead line fault.
- Self-clearing and restarting capability, the first time to realize the use of a hybrid bridge.
- Valve set outputs negative voltage to clear line faults and to be restarted at high speed.
- The world's first hybrid conventional and flexible DC transmission system, with conventional DC at the sending end and flexible DC at the receiving end.
- The world's first hybrid multi-terminal DC transmission project, with a multi-terminal system consisting of conventional DC at the sending end and two flexible DCs at the receiving end.
- Constructed the world's first multi-DC feed-in grid system consisting of flexible DC and conventional DC, with flexible DC providing both active and reactive power to improve grid security and stability.
- Developed the world's first UHV hybrid multi-terminal DC transmission control and protection system, which realized the coordinated control of a multi-terminal system consisting of a conventional DC at the sending end and two flexible DCs at the receiving end, forming the world's most multi-mode DC system.
- Developed the world's largest capacity flexible DC converter (± 800 kV / 5,000 MW) and the world's largest number of flexible DC single-station converter power modules (5,184).
- For the first time in the world, a single valve group and a single station of the UHV hybrid DC system could be cast or canceled online, overcoming the difficulties of DC short-charging and zero-voltage high-current operation of the hybrid bridge valve group.
- The world's first systematic development and manufacture of complete sets of flexible DC equipment with the highest voltage level and largest capacity.
- Constructed the world's largest DC transmission valve hall (89 m long × 86.5 m wide × 43.75 m high).
- For the first time in the world, the stable operation of multi-terminated flexible DC under AC faults has been realized, and full AC fault ride-through has been achieved.
- The world's first long-term reliable operation technology that safely isolates any fault in a single power module.
- For the first time in the world, technical specifications and complete sets of design technologies for UHV conventional DC and flexible DC hybrid transmission technologies have been established.

The first units of Wudongde Hydropower Station were officially put into operation for power generation on June 29, 2020, after 72 hours of trial operation. Wudongde Hydropower Station is an important cascade project in the development of the basin, which is conducive to improving and bringing into play the benefits of the downstream cascades, increasing the guaranteed output and power generation capacity of the downstream step power stations, and promoting local economic and social development and alleviating poverty for the migrant population.

Xiangjiaba Hydropower Station
China's Third-Largest Hydropower Station

Xiangjiaba Hydropower Station is a large hydropower station located on the main stream of the lower Jinsha River in China and is currently the third most completed hydropower station in China and the sixth in the world. The Xiangjiaba Dam is located on the lower reaches of the Jinsha River at the junction of Shuifu City, Zhaotong City, Yunnan Province, and Xuzhou District, Yibin City, Sichuan Province, and is the last stage of the Jinsha River Hydropower Base. The ± 800 kV DC UHV localization demonstration project from Xiangjiaba Hydropower Station to Shanghai is one of the highest and most advanced power systems in China in terms of transmission voltage level.

Xiangjiaba Hydropower Station panorama

The construction of Xiangjiaba Hydropower Station started in November 2006, and the first unit started to generate electricity on November 5, 2012. The hub of the hydropower station is composed of a dam, a flood discharge and sand discharge building, a plant behind the dam on the left bank, an underground plant on the right bank, a vertical ship lift on the left bank, and an irrigation water intake outlet on both sides. The dam is a concrete gravity dam with a crest elevation of 384 m, a crest height of 162 m, and a crest length of 909 m. The powerhouse is located underground on the right bank and behind the dam on the left bank, with four generators in each plant. Since 2013, the right bank behind the dam power station has been under construction, and three generators have been added. Thus, the total number of generators at Xiangjiaba Hydropower Station has reached 11, with a total installed capacity of 6.4 million kW and an annual power generation capacity of 30.747 billion kW·h. Xiangjiaba Hydropower Station is a super dam among 25 hydropower stations in the Jinsha River Hydropower Base that takes into account irrigation functions, while the remaining 24 dams have no irrigation facilities. The scale of the Xiangjiaba ship hoist is comparable to that of Three Gorges, which is the largest single ship hoist in the world. It takes only 15 minutes for a 1,000-ton ship to cross the dam, while the average time for crossing the dam at Three Gorges Locks is 5 hours, and the efficiency of the ship turning over the dam far exceeds that of Three Gorges five-stage locks.

The development task of Xiangjiaba Hydropower Station is mainly to generate electricity, taking into account flood control, navigation, irrigation, sand control, and counter-regulation for Xiluodu Hydropower Station. The power station mainly supplies electricity to central and eastern China, taking into account the electricity needs of the Sichuan and Yunnan provinces.

Xiangjiaba Hydropower Station has created a number of the world's best.

- To overcome the loose geological conditions, Xiangjiaba has created the world's largest sinkhole complex.
- The world's largest 800,000 kW super turbine was installed at Xiangjiaba in order to effectively utilize the huge waterfall of the Jinsha River.
- To match the location advantage of Shuifu, the first port on the 10,000 miles Yangtze River, and not to affect the upstream shipping in Shuifu, the world's largest single-boat lift was built at Xiangjiaba.
- Due to the closeness between Xiangjiaba and urban construction, and the loose geological conditions of the Jinsha River, Xiangjiaba could not use the simple pick-and-pull method of energy dissipation like the Three Gorges Dam, but was forced to choose the method of underflow energy dissipation that is more technically difficult and costs more, for which two of the world's largest large flood dissipation ponds were built.
- The cable machine used is the first giant domestic cable machine with the largest span in Asia.
- The world's longest sand and gravel aggregate conveyor belt, more than 30 km long, was specially built to solve the problem of supplying sand and gravel aggregate to Xiangjiaba.

When the upstream is regulated by Jinping and Xiluodu hydropower stations, the guaranteed output of Xiangjiaba hydropower station is 2.009 million kW, with an annual power generation capacity of 30.747 billion kW·h. In the long term, after the completion of the planned large storage reservoirs in the main upstream tributaries, such as Tiger Leaping Gorge, Lianghekou, and Baihetan, the guaranteed output will increase to more than 3.5 million kW. The power generation capacity and power quality will be steadily improved. The joint operation of Xiangjiaba Hydropower Station and Xiluodu Hydropower Station is one of the main engineering measures to solve the flood control problem of the Yangtze River, and together with other measures, the flood control capacity of Yibin, Luzhou, Chongqing, and other cities can gradually reach the national standard. Meanwhile, together with the Three Gorges Reservoir, the flood control capacity of the Jingjiang River section can be further improved, and flood losses can be reduced in the middle and lower reaches of the Yangtze River. In addition, since the Jinsha River is a mountainous river with narrow channels, many beaches, and rapid flow, the Xiangjiaba reservoir area will become a safe deep-water navigation area for boats, and shipping conditions can be fundamentally improved. Joint scheduling and operation with the Xiluodu reservoir can also enhance downstream shipping conditions during the dry period.

HONGJIADU HYDROPOWER STATION
The Kick-Off of the Great Western Development

Hongjiadu Hydropower Station is located on the main stream of Wujiang River at the junction of Qianxi County and Zhijin County in northwestern Guizhou, and is the only leading power station with a multi-year regulating capacity of water volume among the 11 cascade power stations in Wujiang Hydropower Base. The height of the dam is 179.5 m, and the control basin above the dam site covers 9,900 km², with an average annual runoff of 4.89 billion m³. The total reservoir capacity is 4.947 billion m³, and the regulating reservoir capacity is 3.361 billion m³. Three vertical shaft mixed-flow turbine generator sets are installed in the power station with a total installed capacity of 600,000 kW. The total project investment is 4.927 billion yuan. Hongjiadu Hydropower Station is one of the first key projects to be started for the development of the western part of the country and the "West-East Transmission of Electricity." Due to the regulating effect of the reservoir, it can increase the annual power generation capacity of Suofengying, Dongfeng, and Wujiangdu power stations by 1.596 billion kW·h. After the completion of the power station, it will mainly undertake the tasks of peak regulation, frequency regulation, and accident backup in the Guizhou power grid, which will effectively improve the operating conditions of the power grid.

Hongjiadu Hydropower Station panorama

Hongjiadu hydropower station hub consists of a concrete panel rockfill dam, left bank flood relief system (including cave spillway, flood relief cavern), left bank water diversion power generation system, behind-the-dam-bank type power plant house, and other components. The project officially started on November 8, 2000, and the first unit was put into operation on July 18, 2004, and the 2nd and 3rd units on November 6 and December 20 of the same year, achieving the miracle of "One Year

Hongjiadu Reservoir

and Three Investments" in the construction of hydropower projects of the same scale. The construction period from the start of the diversion hole to the first unit of power generation was four years and three months, and the main project was completed at the end of 2005. The total construction period was shortened by two years and three months compared with the scheduled length of time, which contributes to the grand goal of the West-East Power Transmission Project.

"West-East Power Transmission Project" refers to the development of electricity resources in western provinces (regions) such as Guizhou, Yunnan, Guangxi, Sichuan, Inner Mongolia, Shanxi, and Shaanxi, and their transmission to Guangdong, Shanghai, Jiangsu, Zhejiang, and the Beijing-Tianjin-Tangshan region, where electricity is in short supply. The West-East Power Transmission Project is one of the landmark projects of China's western development. Zhu Rongji, the ex-Premier of the State Council, said in his congratulatory telegram on the start of the project: The start of the West-East Power Transmission Project of Hongjiadu Hydropower Station marked the beginning of the development of western China. Hongjiadu Power Station is mainly responsible for peak regulation, frequency regulation, and standby tasks in the Guizhou power grid, improving power grid operation conditions, and has comprehensive benefits such as flood control and improvement of industrial and agricultural water supply, ecological environment, tourism, aquaculture, and shipping. The construction of the power station will play an important role in promoting the development of the Wujiang River, the optimization of Guizhou's power structure, and the revitalization of Guizhou's economy. In order to build a first-class power station with high efficiency, high quality, economic development, and environmental protection, a lot of new technologies, new materials, and new techniques were adopted, and many engineering technical problems were solved. Ten projects in the construction process won the first prize in Guizhou Province for quality engineering.

Since the commissioning of Hongjiadu Power Station, with safety production as the top priority, by exploring the management mode of "new plant and new system," remarkable results were achieved, easing the coal and electricity tension, and making positive contributions to the "West-East Power Transmission" and "enriching people's prosperity in Guizhou."

GONGBOXIA HYDROPOWER STATION

The First Concrete Panel Rockfill Dam on the Main Steam
of the Yellow River

Gongboxia Hydropower Station is located on the main stream of the Yellow River at the junction of Zunhua Salar Autonomous County and Hualong Hui Autonomous County in Qinghai Province, 153 km away from Xining, and is a large step power station in the upper reaches of the Yellow River. The pivot building consists of dam, water diversion and power generation system, and water discharge system, with a total reservoir capacity of 0.62 billion m³, taking into account irrigation and water supply. Gongboxia Hydropower Station is the first pearl of the northern channel of the West-East Power Transmission Project, in response to the Great Western Development, and is also the first large hydropower station invested in and built after the establishment of the Upper Yellow River Hydropower Development Co. The power station is equipped with five 300,000 kW hydro-generating units, with a total installed capacity of 1.5 million kW and an average multi-year power generation capacity of 5.14 billion kW·h.

The project officially started construction on August 8, 2001, and the first 300,000 kW unit was connected to the grid on September 23, 2004, marking China's installed hydropower capacity exceeding 100 million kW·h. The second unit was put into operation in October 2004, the third unit in July 2005,

Gongboxia Hydropower Station panorama

and the fourth unit on December 5, 2005, after completing a 72-hour trial run. On January 21, 2006, it was connected to the grid after its shortcoming elimination. And on July 5, 2007, all five units were in operation. The commissioning of Gongboxia Hydropower Station has relieved the pressure of electricity consumption in Qinghai Province, met the local power market supply demand, strongly promoted the social and economic development of the region, and wrote a magnificent and strong stroke for the national power supply construction.

The water-retaining dam of the project is the first large reinforced concrete panel rockfill dam project built on the main stream of the Yellow River, with complex geological conditions at the dam site and high-quality requirements for dam construction. The project adopted a number of new technologies, such as self-reinforced concrete extruded side walls and K30 quality inspection for the first time in China to ensure the construction quality of the dam; ten new technologies promoted by the Ministry of Construction, such as EP new insulation materials, optimized concrete mix ratio, and new formwork were also adopted in the construction, which solved the key problems of concrete construction quality control in dry alpine areas and ensured the intrinsic and appearance quality of concrete. For the first time at home and abroad, the drainage building adopted horizontal cyclone energy dissipation technology and successfully converted the diversion cavern into a flood relief cavern, solving the problem of flood relief safety under complex and harsh geological conditions. The reinforced concrete seepage control panel of the dam was poured to the top of the dam at one time, with a sloping length of 218 m, making it the longest concrete seepage control panel in the world at that time. At the same time, various measures such as deep curtain grouting and multi-layer seepage control treatment of construction joints were also adopted to successfully solve the problems of building foundation stability and of leakage of a reinforced concrete panel rockfill dam.

As one of the 13 large pilot projects in China to carry out environmental supervision, The Gongboxia Project has successfully implemented environmental protection and monitoring management, significantly improving the regional ecological environment and opening up a new way for environmental protection in similar water conservancy and hydropower projects in China. The first fish breeding and releasing station was built on the main stream of the Yellow River, pioneering the protection of rare fish species in the river basin. All the sewage from the living area was treated and used for greening and irrigation, achieving "zero discharge" of domestic sewage. Large-scale landscaping is carried out and becomes the first garden-style construction site among projects of similar scale in China's hydropower industry. Thanks to its new concept in hydropower development and construction, Gongboxia Hydropower Station has won almost all the honors in hydropower construction: China Electric Power Engineering Quality Award, Luban Award, the highest award in China's construction engineering, and National Environment-Friendly Engineering Award, the highest award in China's environmental protection engineering. In 2009, the Gongboxia Hydropower Station project won the Zhan Tianyou Award, the highest national award in the civil engineering category.

FENGMAN HYDROPOWER STATION

Parent Hydropower Station in China

Located on the Songhua River in Jilin City, Jilin Province, the Fengman Hydropower Station was built in 1937 and was the largest hydropower station and the most important water hub in the northeast region at the beginning of the founding of the People's Republic of China. Fengman Hydropower Station has gone through lots of ups and downs in its history. The construction of Fengman Hydropower Station started in 1937 during the Japanese invasion of Northeast China, and was the largest hydropower station in Asia at that time. In 1942, the dam began water storage; on May 29, 1943, the first unit was put into operation to generate electricity. On August 10, 1945, Japan announced its unconditional surrender. By then Fengman power plant had not been completed. With the total investment of 0.237 billion yen, 50% of the installation of power station units of the first phase of the project and 87% of the total project volume had been completed. After the PLA recaptured the Northeast from the Kuomintang in 1948, it commissioned

Old dam site of Fengman Hydropower Station panorama

the Soviet Petrograd Institute of Hydropower Design to make the design for the restoration and expansion of the Fengman Hydropower Station, installing one 60,000 kW, two 65,000 kW, five 72,500 kW, and one 1,250 kW unit sets, the total installed capacity being 0.55375 million kW. Later, three more phases of rehabilitation and expansion projects were carried out.

The old dam of Fengman Hydropower Station is 91 m high, 1,080 m long, 60 m wide at the bottom, and 9–13.5 m wide at the top, with a concrete volume of 1.94 million m³. The design flood discharge is 9,020 m³/s, and the design flood level is 266 m, forming a reservoir with a storage area of 550 km² and a total storage capacity of 10.78 billion m³. In 1942, when 59% of the dam concrete pouring was completed, water began to be stored. On October 17 of that year, the interception gates began to be closed, and on November 7, the last one was closed, and thus the interception was completed. The ancient flow of the west flow of the Songhua River (then called the "second Songhua River") was cut off, and from then on emerged the man-made Songhua Lake. As the Japanese invaders expanded their front, especially after the Japanese attack on Pearl Harbor, the United States joined the war, which led to the Pacific War. The shortage of resources in Japan became more and more obvious, and the Japanese regime was eager to complete the power generation as soon as possible, speeding up the project at all costs, and using a large amount of inferior concrete and other construction materials. The construction procedures were even more chaotic, resulting in serious defects in the main project of the dam, with water leakage and other problems. Even so, when Japan announced its unconditional surrender in 1945, the concrete pouring of the Fengman Dam was only 89% of the total concrete, and the height of the dam had not yet reached the initial design height. During the period from the surrender of Japan to the founding of the People's Republic of China, the Kuomintang rulers did not carry out any reinforcement, repair, or further construction of the Fengman Hydropower Station, causing the dam to continue to deteriorate. After the founding of the People's Republic of China, the Fengman Hydropower Station was finally given back to China. In order to complete the project as soon as possible so that it could perform its original design function and contribute to China's economic and social development, the Chinese government made the Fengman Hydropower Station a key project in the "First Five-year Plan." Under the leadership of the Party and the State, the builders made concerted efforts to build the Fengman Hydropower Station, which was basically completed in 1953.

Fengman Hydropower Station is a backbone power station in the northeast grid, providing not only a large amount of electricity, but also playing an important role in peak regulation, frequency regulation, and accident backup, with one unit regularly generating electricity as base load, maintaining a certain basic flow of the west-flowing Songhua River to meet shipping requirements, and other units mainly for peak regulation in the grid. The other units are mainly used for power grid peaking. The power station also has comprehensive utilization benefits such as flood control, water supply, irrigation, aquaculture, and tourism. As an old hydropower plant, Fengman Hydropower Station has accumulated a wealth of valuable experience in long-term production and has trained and provided a large number of talents for China's hydropower industry. The development history of Fengman Hydropower Station also tells us that we will be beaten if we fall behind, and we must develop in order to strengthen ourselves!

The old dam of Fengman Hydropower Station is currently the longest-operating large concrete gravity dam in China. The construction has been through the Japanese Puppet Regime till after the establishment of the People's Republic of China. Due to historical conditions, the construction quality is poor: There are dam leakage, dissolution, concrete aging freeze-thaw damage, poor integrity of the dam, and many other problems and congenital defects, despite years of reinforcement and maintenance. The State Electricity Regulatory Commission (SEDC) rated the Fengman dam as dangerous in 2007, and in 2009 its owner, the State Grid Corporation (SGC), determined the reconstruction program. On October 18, 2012, the National Development and Reform Commission approved a comprehensive treatment (reconstruction) project for the Fengman Hydropower Station. The reconstruction project is to build a new dam 120 m downstream of the original dam. The dam, which restores the original power station functions and tasks without changing the characteristic reservoir level of water, is the world's first large hydropower station reconstruction project with nearly 100 m dam height, 10 billion reservoir capacity, and millions of installed machines. The new dam is a compact concrete gravity dam with a length of 1,068 m and a maximum height of 94.5 m. The normal reservoir storage level is 263.5 m, with a multi-year regulation capacity. The power station will install six new single-unit 200,000 kW Francis turbine generators and retain two 140,000 kW units from the original Phase III project, with a total installed capacity of 1.48 million kW, which will have an average annual power generation capacity of 1.709 billion kW·h and will be connected to the Jilin power grid at 500 kV.

Construction of the new dam at Fengman Hydropower Station

On December 12, 2018, the old Fengman dam underwent its first blasting. The length of the dam involved was 18 m, the width was 13.5 m, and the height was 4.2 m. This also marked the end of the old dam of Fengman Hydropower Station, operated for 80 years, and the end of its historic mission. The original dam of Fengman Hydropower Station was successfully blasted at 11:16 on May 20, 2019. Now the new dam has been completed, and the demolition work of the old dam has been finished. The new dam has twice the flood control capacity and 15 times the power generation capacity of the old one, and the major safety hazards that have plagued people for years have been completely eliminated. At present, the new Fengman dam is of excellent quality, with a dam base leakage of 7.7 L/s, reaching the leading international level. Meanwhile, the new Fengman Hydropower Station has restored the flood level of the reservoir to the normal flood level of 263.5 m elevation, increasing the maximum flood control capacity by one time and completely eliminating the defects of the old hydropower station.

Blasting scene of the old Fengman Dam

The construction process of the old Fengman Hydropower Station for the development of the national hydropower industry has accumulated a wealth of operating experience, training a large number of excellent backbone experts and technical personnel. According to estimated statistics, the Fengman Hydropower Station has come out of a total of more than 2,000 technical personnel to support the construction of water conservancy projects in other parts of the country, who later appeared in hydroelectric projects, including Liujiaxia, Longyangxia, Gezhouba, Yangtze River Three Gorges … Fengman Hydropower Station is thus entitled "Parent Hydropower Station in China."

GUTIAN CREEK HYDROPOWER STATION
China's First Concrete Wide-Joint Gravity Dam

Gutian Creek Hydropower Station is the earliest developed cascade hydropower station in China, designed, constructed, manufactured, and installed by China independently, and it is a typical epitome of hydropower construction in the early years of the People's Republic of China. Gutian-I Hydropower Station is located on Gutian Creek in Gutian County, Fujian Province, with a 1,920 m-long diversion tunnel, a 4.4 m-diameter hole, and a 59.6 m-long, 12.5 m-wide, 29.5 m-high underground plant. It is also China's first underground power plant, with two sets of 6,000 kW and four sets of 12,500 kW hydro-generating units installed. The first 6,000 kW unit was put into operation in March 1956.

Fujian Gutian Creek Hydroelectric Power Plant is a state-owned medium-sized power generation enterprise with one plant and four stations under centralized management. The plant consists of 4 power stations (12 units) with a total installed capacity of 259,000 kW, which has reached 276,000 kW in recent years after capacity increase and renovation. The designed annual average power generation is 0.741 billion kW·h. It is responsible for peak regulation, frequency regulation, voltage regulation, and accident backup. The plant is located in Gutian County, 6 km away from Gutian County, and its sites and

Gutian First Stage Hydropower Station panorama

stations are located in Gutian and Minqing counties, stretching nearly 40 km. As the saying goes, "Among the renowned Hydropower Stations in China, Fengman is counted as representative of the northern China and Gutian of southern China." Gutian Creek Hydroelectric Power Plant is the first underground hydroelectric power plant built in new China. In March 1951, the construction of the first phase of the underground plant started. In March 1956, two units of the first phase of the plant were put into operation, and by December 1973, all 112 units of each cascade of the plant were put into operation.

The first and second cascades of the Gutian Creek Step Hydropower Station are in Gutian County, while the third and fourth are in Minqing County.

(1) The first cascade power station is a tunnel-diversion type power station located 5 km away from the new city at Bankengting. Construction started in March 1951, and the first stage was completed in 1956 with a temporary timber frame rockfill dam, 176.5 m long and 6 m high, intercepting the entire low water riverbed. The underground plant is 83 m long, 12.5 m wide, and 29.5 m high, with a capacity of 6 units, and the first phase installed two units of 120,000 kW. On March 1, 1956, it was put into operation to generate electricity for Fuzhou City. Construction of the second phase started in January 1957 and was completed in August 1960. A 412 m-long, 71 m-high concrete wide-slit gravity dam was built at Turtle Mountain to form the Gutian Reservoir. The second phase was increased to six units with an installed capacity of 620,000 kW. Power was generated in August 1960 to supply Fuzhou, Nanping, and Sanming. The underground construction of this cascade of the power station also includes the diversion tunnel, which is 1,758 m long and has an internal diameter of 4.4 m; the regulating well is 80 m deep and 11.9 m in diameter.

(2) The second cascade power station is a tunnel-diversion type power station, the dam site is in Longting, and the power plant is located in Houyang, Minqing County. The construction started in 1958 and ended in 1969. The dam is an overflow-reinforced concrete flat dam, 208.5 m long, with a maximum dam height of 43.5 m. The diversion tunnel is 5,249 m long and has an inner diameter of 6.4 m. The diversion ground plant is 51 m long, 19.6 m wide, and 36.9 m high, with two installed units (130,000 kW) and an average annual power generation capacity of 0.347 billion kW·h for many years, which is the largest power station among the cascaded power stations.

(3) The third cascade power station is a back-of-dam power station located in Gao Yang, Minqing County. The construction started in September 1958, and two units with a capacity of 330,000 kW were installed from 1965 to 1973, with an average annual power generation capacity of 0.114 billion kW·h for many years.

(4) The fourth cascade power station is a back-of-dam power station located in Baohu, Minqing County. The construction started in 1965, and two units with a capacity of 340,000 kW were installed successively from 1971 to 1972, with an average annual power generation capacity of 0.124 billion kW·h for many years.

In the early 1950s, when the country was in a state of need, the central government listed the Gutian Creek cascaded Hydropower Station as the "101st" key construction project in the country when formulating the "First Five-year Plan." The development and construction of Gutian Creek Hydropower Station took more than 20 years, making it the longest hydropower station construction period in the history of China. It has not only accumulated rich experience for China in constructing cascaded hydropower station, but also trained a large number of hydropower construction and management talents for China, Vietnam, and some African countries. So it is known as the cradle of hydropower talents in new China.

MANWAN HYDROPOWER STATION
China's First Million-Kilowatt Hydropower Station Built as Joint Venture by Central and Local Governments

Manwan Hydropower Station is located on the midstream section of the Lancang River 1 km downstream of the mouth of Manwan River at the junction of Yunxian and Jingdong counties in western Yunnan Province, China, 140 km from Lincang and 200 km from Dali City. The hydropower station is developed with the single objective of power generation. The Manwan Barrage is a concrete gravity dam with a height of 132 m and a total reservoir capacity of 0.92 billion m³. Construction of the power station officially started on May 1, 1986; the river was cut off in December 1987, the first unit was connected to the grid in June 1993, and all five units were put into operation in June 1995, and by then the first phase of the project was basically completed. After the upstream construction of Xiaowan Hydropower Station, the second phase of this power plant will have an installed capacity of 250,000 kW, with a total installed capacity of 1.5 million kW, a guaranteed output of 0.796 million kW, and an annual power generation capacity of 7.88 billion kW·h.

Manwan Hydropower Station panorama

Manwan dam site controls a basin area of 114,500 km², a multi-year average flow of 1,230 m³/s, a normal storage level of 994 m, a dead water level of 982 m, extraordinary flood level of 997.5 m, a total reservoir capacity of 0.92 billion m³, regulating reservoir capacity of 0.258 billion m³, for the seasonal balancing reservoir. The reservoir area is 23.9 km², the design flood flow is 18,500 m³/s in 100-year return period floods, the calibration flood flow is 22,300 m³/s in a 5,000-year return period, and the possible maximum flood flow is 25,100 m³/s. The multi-year average sand transport is 40 million t, the measured maximum sand content is 14.3 kg/ m³, and the average sand content is 1 kg/m³.

Manwan Hydropower Station is mainly composed of a barrage dam, power plant, water discharge buildings, and so on. Barrage dam top elevation 1,002 m, dam top length 418 m. It is divided into 19 dam sections. Among these, dam section 1–7 are for non-overflow dam section; 1 sand flushing bottom hole with an internal diameter of 6 m are arranged for dam section 8 and 14, respectively; dam section 9–13 are for overflow dam section, each with a 7.5 m inner diameter steel pipe for power generation at 945 m elevation and five 13 m × 20 m (width × height) overflow table holes at 974 m elevation; the dam section 15 has two 5 m × 8 m (width × height) discharge bottom holes at 925 m elevation; the dam sections 16 to 19 are non-overflow dam sections.

The plant is of the overflow type behind the dam, with a fully enclosed structure. The total length of the main plant is 195 m, the height is 59.9 m, the net width is 34.5 m, the top arch sagittal height is 3.13 m, the top arch thickness is 4 m, and the unit spacing is 26 m. Six units with a single capacity of 250,000 kW are installed, and the turbine type is vertical shaft mixed-flow type, of which five units are installed in the first phase of the project, with four 220 kV and three 500 kV transmission lines, and one additional 250,000 kW unit is installed in the second phase of the project. The maximum head of the unit is 100 m, the design head is 88.12 m, the rated head is 89 m, and the minimum head is 69.3 m. 220 kV substation and 500 kV substation are overlapped on the top of the left and right installation rooms, respectively, and both are indoor types.

Manwan power station's flood discharge is dominated by 5 table holes at the top of the dam, supplemented by the left bank flood tunnel and left bank flood discharge double middle holes, left and right bank sand drainage bottom holes (drainage of the reservoir). The five overflow table holes are at the top of the overflow dam, each with a 13 m × 20 m (width × height) arc gate, the top of the weir elevation 974 m, double-layer differential diffusion picking flow with high kink (35°) for No. 1, No. 3 and No. 5 and low kink (23°) for No. 2 and No. 4, and the water pad pond energy dissipation is set under the dam. The flood relief hole is located in the upper part of the No. 1 diversion hole within the mountain on the left bank, and is arranged in a zigzag pattern with No.1 and No. 2 diversion holes. The inlet elevation of the flood relief cave is 965.5 m, which is composed of wall type inlet chamber, pressurized tunnel, non-pressurized tunnel, outlet anti-arc section, and curved angle-shaped pick flow nosecone. The inlet chamber is 41.72 m long and 19 m wide, with a 12 m × 135 m (width × height) flat maintenance door and a 12 m × 12 m (width × height) arc-shaped working door. In front of the arc-shaped working door is a pressurized inlet with a control size of 12 m × 12 m (width × height). The cavern is a city gate-shaped unpressurized tunnel, with a length of 304.98 m and a net section size of 12 m in width and 15.5 m in height. The outlet section is 67.28 m long (including the open arch section, anti-arc section, and curved angle pick flow inclined nose section). The designed flood discharge volume of the flood relief cave is

2,310 m³/s, and the maximum flood discharge volume is 2,560 m³/s. The flood discharge volume is 16,805 m³/s when each flood relief building is fully opened.

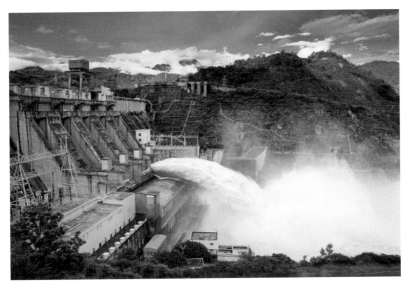

Manwan Hydropower Station flood release

The installed capacity of Manwan Power Station accounts for 1/3 of the capacity of the entire Yunnan power grid system, greatly relieving the tension of Yunnan's electricity, playing a very important role in promoting Yunnan's economic take-off and the prosperity and stability of the frontier, opening up an era of prosperity for Yunnan, and promoting the "west-to-east" and "Yunnan-to-outside" power supply. The project has played a very important role in opening up the era of promoting Yunnan's economic growth and the prosperity and stability of the frontier. The National Acceptance Committee spoke highly of the Manwan project: "The construction period of Manwan Hydropower Station is short, the investment is provincial, the quality is good, and the economic benefits are great, setting six records in hydropower construction nationwide and ranking among 'the five golden flowers' of hydropower nationwide."

WUQIANGXI HYDROPOWER STATION

Hydropower Station in Hunan Province with the Largest Installed Capacity

Wuqiangxi Hydropower Station in Hunan Province has a total installed capacity of 1.2 million kW (5 × 240 MW) and is the backbone power station of the Yuanshui Basin hydropower gradient development and the backbone peaking and frequency regulation power station of the Central China power grid. Located on the main stream of Yuanshui in Yuanling County, which controls 93% of the Yuanshui basin area, Wuqiangxi Hydropower Station started to be surveyed in 1952. The construction started and halted twice in 1956 and 1980 and resumed in April 1986. The first unit generated electricity in December 1994, and all units were put into operation at the end of 1996. With a total investment of 8.9 billion yuan, the project has introduced advanced technology and equipment from 11 countries, including the United States, Japan, and Germany, and is a key project of the National "Seventh Five-Year Plan" and the "Eighth Five-Year Plan."

Wuqiangxi Hydropower Station is mainly for power generation, and also has comprehensive benefits such as flood control and shipping. The guaranteed output of the unit is 255 MW, with a designed multi-year average power generation capacity of 5.37 billion kW·h. Two 500 kV lines are used, one for Loudi

Wuqiangxi Hydropower Station panorama (I)

Minfeng and the other for Changde Gangcheng, forming the first 500 kV main ring network of the Hunan power grid. The power station adopts the dam-back type plant, and the dam is a concrete gravity dam. The maximum dam height is 85.83 m, the normal storage level is 108 m, and the total reservoir capacity is 4.29 billion m³. It is a seasonal reservoir. With the cooperation of the upstream Fengtan Reservoir, the flood control standard of the coccyx dikes can be increased from a 5-year return period to a 20-year return period. The navigation building is a single-line continuous three-stage ship lock with a designed annual freight capacity of 2.5 million t and an annual water crossing capacity of 460,000 m³.

Wuqiangxi Hydropower Station panorama (II)

The project consists of a concrete gravity dam, plant, spillway building, and three-stage ship lock. The dam top elevation is 117.5 m, and the total length of the dam top is 724.4 m. The overflow dam is located on the left side of the river bed in the main channel, with nine table holes, one middle hole, and five bottom holes. The top elevation of the weir of the table hole is 87.8 m, and each hole is 19 m wide and 23.3 m high, with a design discharge capacity of 39,988 m³/s and a maximum flow rate of 50,522 m³/s. The elevation of the bottom of the middle hole is 76 m, the width of the orifice is 9 m, the height is 13 m, with the maximum discharge flow is 2,586 m³/s. The bottom hole is arranged in the gate pier, with an inlet bottom elevation of 67 m, a width of 3.5 m, a height of 7 m, with a maximum discharge flow rate of 3,015 m³/s.

On the right side of the riverbed is a post-dam type plant, 251 m in length, 36.5 m in width, and 68 m in height, which houses five single turbine generator sets with 240,000 kW of mixed flow capacity. The turbine has a runner diameter of 8.6 m, maximum head of 60.1 m, minimum head of 36.2 m, rated head of 44.5 m, rated speed of 68.2 r/min, rated power of 248,000 kW, and maximum efficiency of 0.95. The rated capacity of the generator is 266.67 MVA with a rated power factor of 0.9 and a rated voltage of 15.75 kV, and it is connected to the Central China Grid at 500 kV.

The 1,028 km-long Yuan River rushes from the gorge southeast of Guizhou, rolls eastward through western Hunan, and pounces to the vast Dongting Lake. Among the four major water systems in Hunan, namely Xiang, Zi, Yuan, and Li, Yuan is especially abundant and turbulent, with a drop of 1,033 m containing more than 7 million kW of water energy. After the founding of the People's Republic of China, Hunan has built hydroelectric power stations such as Tuoxi, Fengtan, Dongjiang, and some thermal power plants, but overall the number and the scale are small and cannot meet the needs of its economic and social development. It has been the expectation of several generations of Hunan people to construct a large hydropower station in the main stream of the Yuan River. In April 1986, the State Council approved the resumption of construction of Wuqiangxi Hydropower Station and listed it as a key project in the national "Seventh Five-Year Plan." The geological and hydrological survey started in 1952, and the resumption of construction started in 1986. A total of 120,000 people migrated to other places. In 1996, all five units were put into operation. It is the first large hydropower project with 1.2 million kW in Hunan Province, creating a feat of hydropower construction and setting up a monument in the history of hydropower construction in Hunan.

Construction layout of Wuqiangxi Hydropower Station

The Wuqiangxi project took ten years to complete, and at that time, it boasted ten "national bests," such as successfully managing the highest slope of the domestic hydropower station, adopting the largest table hole arc gate, and the largest mixed-flow stainless steel turbine runner, and building the largest 3-stage ship lock in China. After the completion of the Wuqiangxi hydropower station hub project, the lake reservoir area of 170 km² is formed and becomes a real high gorge flat lake. Five streams rushing out from the mountains of western Hunan converge here, forming vast misty waves, while small and large hills turned into small islands with different postures, so Wuxi Lake (five streams lake) becomes a high mountain lake with a thousand islands. It forms an artificial lake group together with the adjacent Mingyue Lake in the Lingjintan reservoir area, Meng Lake in the Yanwutan reservoir area, Mingxi Lake in the Gaotan reservoir area, and Mangdong Lake in the Fengtan reservoir area. In October 1994, the state approved the establishment of Hunan Wuling Hydropower Development Co., Ltd. with Wuqiangxi Hydropower Station and Lingjintan Hydropower Station as the parent companies, and authorized it to be fully responsible for the scrolling development of the Yuanshui Basin. Wuqiangxi, Lingjintan, Hongjiang, Wumipo, Sambanxi, Tuokou … one by one, many hydropower stations have been built in the main tributaries, upper and middle reaches of the Yuan River, echoing head and tail and complementing each other. The channel of the Yuan River has been totally canalized, and the rapids have been turned into high-gorge flat lakes. Thousand-ton ships can reach Dongting Lake directly from the southwest part of Guizhou. Thousands of miles of Yuan River has become a green energy corridor and transportation corridor, enriching people's life! Over the past 20 years since its completion, Wuqiangxi Hydropower Station has generated an annual average of 5.37 billion kW·h of electricity, constantly sending clean energy to Hunan Province; it has mitigated flood disasters for the Dongting Lake area and even the middle and lower reaches of the Yangtze River as a key reservoir for national flood control. In March 2019, the expansion project of Wuqiangxi Hydropower Station started. Four years later, the installed capacity of the station will reach 1.7 million kW, and the average annual power generation will increase to nearly 6 billion kW·h. The battle of hydropower gradient development in Yuanshui Basin will be perfectly concluded.

SHUIKOU HYDROPOWER STATION
Top of the "Five Golden Flowers"

Shuikou Hydropower Station is located on the main stream of Minjiang River in Minqing County, Fujian Province, 94 km upstream from Nanping City, 14 km downstream from Minqing County, 84 km from Fuzhou City, 112 km from Changmen Inlet and 118 km from Meihua Inlet. It has a total installed capacity of 1.4 million kW and a design average multi-year power generation capacity of 4.95 billion kW·h. It also has comprehensive utilization benefits such as shipping, wood passing, and flood control, and undertakes tasks such as peak regulation, frequency regulation, and accident backup for the power grid.

Shuikou Hydropower Station flood gates opening and discharge photo

As a World Bank loan construction project, Shuikou Hydropower Station was listed as a national capital construction project, and pre-construction preparations were started in 1985. The main body of the project was officially started on March 9, 1987. The Fujian Provincial Electric Power Bureau set up the Shuikou Engineering Construction Company to manage the construction, and the East China Survey and Design Institute undertook its engineering design work. Through international bidding, the construction part of the civil works of Shuikou Hydropower Station was decided to be undertaken by Huatian Associates, with Fujian Provincial Electric Power Bureau as the responsible party.

Shuikou Hydropower Station is a first-class project, consisting of a dam, plant, over-dam building, spillway, etc. The dam is a concrete gravity dam with a maximum height of 100 m and a crest length of 783 m. The normal storage level of the reservoir is 65 m, the dead water level is 55 m, and the effective storage capacity is 700 million m³; in the main flood season from April to July, the design flood level is 61 m, and the corresponding hydraulic capacity is only 320 million m³, with poor regulation performance. The left bank is behind the dam-type hydropower plant, equipped with seven axial flow hydroelectric generating units with a single capacity of 200,000 kW. The right bank is the over-dam building, and the spillway is a riverbed type arrangement, with 12 table holes of size 15 m × 22 m and two bottom holes of size 5 m × 8 m. The ship lock is a 3-stage building. Each lock chamber is 160 m long, 12 m wide, and 3 m deep. The ship lift is arranged on the right side of the lock, and the effective size of the shipping compartment is 124 m long, 12 m wide, and 2.5 m deep. The main buildings of the hydropower station are designed according to the 1,000-year return period flood standard and the 10,000-year return period calibrate standard.

Shuikou Hydropower Station construction started in 1987

Shuikou Hydropower Station controls a watershed area of about 52,438 km², accounting for 86% of the total watershed area of Min River. The average annual precipitation in the basin is 1,758 mm, the average multi-year flow at the dam site is 1,728 m³/s, the total annual runoff is 54.5 billion m³, the maximum measured flow is 30,200 m³/s, the minimum flow is 196 m³/s.

Non-soluble magmatic rocks and clastic rocks surround the reservoir area, and permanent seepage problems and major bank instability do not exist. The dam site area is a hard and intact Mesozoic Yanshan-age black mica granite with an average wet compressive strength of 100 MPa. No major tectonic fractures are found in the dam site area except a few small faults with steep dip angles in the riverbed and extrusion crushing zone, weathering crushing zone.

The reform and opening-up policy provided a good environment for Shuikou Hydropower Station. The project builders were able to aim at the world's advanced technology level and tried their best to catch

Shuikou Hydropower Station

up, and the starting point for technological progress in construction was higher than ever. The design used advanced ideas and methods, and the construction introduced advanced devices and technologies. The specialized hydraulic laboratory of the university tested and proved the river flow channel changes, and the world's largest axial runner at that time was tested in the international identification model. The construction site was highly mechanized, with relatively little construction labor, and repeatedly set records for daily concrete pouring volume. Shuikou Hydropower Station project boldly introduced foreign advanced construction equipment, technology, and management mode, making the construction of the power station have high efficiency and quality. It also developed new construction technology and new materials according to China's reality. For example, new technologies and techniques such as geomembrane impermeable walls, compact concrete, magnesium oxide micro-expansion concrete, and slipform pouring were put into the construction of the power station, which made the power station generate huge economic and social benefits.

Shuikou Hydropower Station is a national "Seventh Five-Year Plan" key construction project invested in by the state and is the first energy project built with World Bank loans in Fujian Province. It ranked high in the national quality evaluation of million-kilowatt hydropower stations and was approved to reach international standards. Five million-class hydropower stations in the 20th century are praised as "five golden flowers" ("five golden flowers" refers to Wuqiangxi Hydropower Station, Diheyan Hydropower

Station, Yantan Hydropower Station, Shuikou Hydropower Station, Manwan Hydropower Station). Shuikou is among them and ranks first. With years of operation, the equipment deterioration becomes obvious. In order to ensure safe and reliable operation, improve water energy utilization and increase flood power generation, Fujian Shuikou Power Generation Group Company, working with Shuikou Hydropower Station annual maintenance, carried out capacity increase and renovation of seven units one after another with a progress of one unit per year from 2019 onwards. According to the plan, Shuikou Hydropower Station will complete the capacity increase and renovation of all seven units in 2026, and the total installed capacity will increase from 1.4 million kW to 1.61 million kW, which is 15% more than the original capacity. After preliminary calculations, the capacity renovation will increase the utilization rate of the hydropower plant from 86.2% to 89.4%, effectively reducing the water abandoned by the dam and increasing the annual power generation by about 0.24 billion kW·h. The capacity renovation will effectively ensure safe production and increase the value of state-owned assets. At the same time, it will increase the peaking regulation, frequency regulation, and accident backup capacity of the Fujian power grid. It will significantly improve the water utilization rate and internal economic return rate of the project, and will be of great significance to improve the ecological environment quality of Minjiang River Basin and serve the clean energy development of Fujian Province.

LIJIAXIA HYDROPOWER STATION

The World's Largest Hydropower Station with Two-Row Placed Units

Lijiaxia is located in the middle section of Lijiaxia Valley, the main stream of the Yellow River, at the junction of Jianzha and Hualong counties in Qinghai Province, 1,796 km from the source of the Yellow River and 3,668 km from the mouth of the Yellow River. It is the third-level large-scale hydropower station in the cascade development of the upper reaches of the Yellow River. The construction was formally started in April 1988 and was divided into two phases. The main part of the project is a three-centered arc double-curvature arch dam with a maximum height of 165 m, a crest length of 414.39 m, and a bottom width of 45 m; the crest elevation of the dam is 2,185 m., and the normal reservoir water level is 2,180 m, with a total capacity of 1.65 billion m³. The plant is a double-row arrangement behind the dam, with five units installed in a plum-shaped distribution. The single unit capacity is 400,000 kW·h, and the design annual average power generation capacity is 5.9 billion kW·h.

Lijiaxia Hydropower Station panorama

The construction of the Lijiaxia Hydropower Station was officially started in April 1988, and the flow was cut off on October 13, 1991. The Lijiaxia Hydropower plant was officially established on December 12, 1995, and the reservoir started water storage in December 1996. In December 1999, the Yellow River Upstream Hydropower Development Co., Ltd. was established, and the Lijiaxia Hydropower Plant was renamed the Lijiaxia Power Generation Branch, which was completed in the fourth quarter of 2001. The stator winding of No. 4 hydro generator of Lijiaxia Hydropower Station, which adopts evaporative cooling technology, is listed as a key science and technology project of the National "Ninth Five-Year Plan." As a new cooling technology, evaporative cooling can greatly reduce the temperature and temperature difference of the generator stator winding and the thermal stress of the generator stator during operation, and significantly improve the utilization rate of the material, thus providing a better safety guarantee.

The dam site area has a V-shaped river valley, with the foot of the right bank at about 50° and the foot of the left bank at about 45°, and the width of the river channel is about 50 m. The surrounding rocks of the reservoir are weakly permeable and relatively impermeable, and the normal reservoir storage level is lower than the groundwater divide on both banks. The groundwater flows to the Yellow River, so there is no permanent leakage problem because there is no adjacent valley or drainage lowland below the normal reservoir storage level. The bedrock of the dam site area is mainly a mixture of pre-Auroraic black mica long-striped rocks and black mica hornblende plagioclase. Due to the influence of geological and tectonic movements, faults, fissures, and extrusion zones are more developed in the dam foundation stability analysis and treatment, and the stability, deformation, and treatment of the left and right shoulders of the dam were difficult problems in the engineering design and construction at that time. The power station has taken effective management measures for the deep anti-slip stability of the rock shoulder of the double-curvature arch dam, the large reservoir bank landslide near the dam, and the landslide in the downstream energy dissipation area, and accumulated experience in experimental research, construction and safety monitoring for the deep shoulder treatment of arch dams and comprehensive management of large landslides in China.

As the first large hydropower plant in the upper reaches of the Yellow River to be built by bidding, the diversion tunnel, foundation excavation, cofferdam cut-off, and power station pivot works of Lijiaxia Hydropower Station were all constructed by the Fourth China Water Engineering Bureau, which has experience in building two-megawatt hydropower stations, Liujiaxia and Longyangxia. Since the commissioning of the Lijiaxia Hydropower Station, the water conservators of Lijiaxia have been seeking innovation in management mechanisms and management concepts, and the comprehensive management level has been steadily developing and improving, and the production tasks assigned by the higher level have been completed without slackness. At the same time, through the development of hydropower resources in the upper reaches of the Yellow River, the minority areas in the upper reaches of the Yellow River have been developed very quickly, and the living standards of the people have been improved. The completion and operation of Lijiaxia Hydropower Station have strongly supported the national strategy of western development and made a great contribution to social and economic development.

GOUPITAN HYDROPOWER STATION

Guizhou "West-East Power Transmission" Landmark Project

Located in Yuqing County, Guizhou Province, Goupitan Hydropower Station is the fifth cascade of the Wujiang River hydropower gradient development, 137 km upstream from Wujiangdu Hydropower Station and 455 km downstream from Fuling, the river mouth. 43,250 km² of the basin area is under control, 49% of the Wujiang River basin, with a normal storage level of 630 m, a total reservoir capacity of 6.454 billion m³, and a regulating reservoir capacity of 2.902 billion m³. The dam site has an average multi-year flow of 717 m³/s, and an average multi-year runoff of 22.6 billion m³. Goupitan Power Station is mainly for power generation, while taking into account other tasks such as shipping and flood control. It is equipped with five generating units with a total installed capacity of 3 million kW and a single unit

Goupitan Hydropower Station panorama

capacity of 600,000 kW. The main task of the power station is to generate electricity, taking into account shipping, flood control, and other comprehensive use.

On November 8, 2003, Goupitan Hydropower Station officially started construction, and on November 16, 2004, it achieved the interception of the Grand River, and on July 31, 2009, the first unit was put into operation to generate electricity, and on December 29, 2009, it achieved the great feat of five throws in one year for large domestic units. The power station hub consists of an arch dam, plant, dam body table flood holes, and left bank flood cavern. The dam is a concrete double-curvature arch dam, 225 m high, with an arc length of 553 m, an arc height ratio of 2.38, a crown bottom thickness of 50.28 m, and an underground plant. There are two diversion holes on the left bank and one diversion hole on the right bank, which form the engineering diversion system.

It is a thin arch dam with a height of more than 200 m. It has a high-flow flood discharge and energy dissipation design, a large underground plant cavern group, a large-diameter inflow tunnel through soft rock, and an RCC cofferdam with a height of more than 70 m, all of which are design challenges. The dam base and shoulders are distributed with interlayer misalignment and dissolution zones that are basically parallel to the rock direction, as well as some karst caves, and poor compression and deformation conditions of the dam shoulder rock. In order to reduce the impact of these defects, the following treatment measures were taken: the karst caves within the arch dam holding capacity, which have a greater impact on the stress, deformation, and seepage control requirements of the arch dam, were backfilled with concrete; the interlayer misalignment zone was treated with a lattice-like replacement scheme of flat caverns and inclined wells, which resulted in a symmetrical distribution of stress in the dam body and uniform stress at the arch end, and improved the plastic zone and increased the overload capacity of the arch dam. The safety of the project has been significantly improved.

The maximum head difference between the upstream and downstream of the hydropower station is 150 m, and the flood discharge volume and power of the dam are at the leading level at home and abroad, with the design and calibration flood discharge volume reaching 24,016 m^3/s and 28,807 m^3/s respectively, and the flood discharge power reaching 34,940 MW and 41,690 MW respectively. The topography and geology are complex, so the design of flood discharge and energy dissipation is much more difficult. Through a comprehensive comparison of flood discharge arrangement, energy dissipation effect, and engineering volume, six table holes and seven middle holes were used for flood discharge, and water cushion ponds were set up under the dam to dissipate energy. The top elevation of the weir of the flood relief table hole is 617 m, the size of the orifice is 12 m × 15 m, and the arc gate is installed. In order to disperse the water flow into the pond, the outlet is divided into two types: upward pressure plate type and flat bottom type.

As a key project of the National "Tenth Five-Year Plan," Goupitan Hydropower Station is the largest hydropower station built by Guizhou Province and China Huadian Group Corporation, and embodied West-East Power Transmission Project in Guizhou.

JILINTAI-I HYDROPOWER STATION
The 10th Cascade Hydropower Station in the Kashgar River Basin Plan

Jilintai-I Hydropower Station is located on the Kashi River in Hujiltai Township, Nilek County, Xinjiang Uygur Autonomous Region, 32 km from Nilek County and 142 km from Yining City, with a total installed capacity of 500 MW, mainly for power generation, irrigation and flood control. Above the dam site, it controls a watershed area of 6,163 km², a total reservoir capacity of 2.53 billion m³, regulating reservoir capacity of 1.7 billion m³ for an incomplete annual balancing reservoir. Four 115,000 kW mixed-flow turbine generator sets are installed, with a total installed capacity of 460,000 kW and a multi-year average power generation capacity of 0.938 billion kW·h. The total project investment is RMB 2.286 billion.

On May 1, 2001, the construction of the diversion cavern started. On September 15, 2002, the power station cut off the flow, and in October 2004, the water was stored in the gate. In July 2005, the first unit was connected to the grid. All units were connected to the grid in April 2006, and the project was completed in October 2006.

Upstream of Jilintai-I Hydropower Station

The normal storage level of the reservoir is 1,420 m, the dead water level is 1,380 m, the flood limit level is 1,419.5 m, the design flood level is 1,420.05 m, and the calibration flood level is 1,422.19 m. The building consists of a concrete panel gravel dam, table hole flood cavern, deep hole flood cavern, power generation tunnel, pressure pipeline, and power plant room. The top elevation of the dam is 1,425.8 m, the maximum dam height is 157 m, and the top length is 445 m. The slope of the upstream dam is 1:1.7, the slope of the dam between the downstream packway is 1:1.5, and the average slope of the dam is 1:1.96. The dam material includes riverbed gravel in the upstream part of the dam axis and blasting excavation material in the downstream part. The total amount of earth and rock excavation for the dam is 1.968 million m³, the total amount of filling is 9.445 million m³, and the concrete is 71,400 m³.

The table hole flood cavern is arranged in the mountain body on the left bank, with a section of 7 m × 8.5 m in the shape of a city gate cave, the tunnel length is 773 m, the inlet weir top elevation is 1,411 m, the maximum discharge capacity is 581.5 m³/s, using the pick flow to dissipate energy. The deep hole flood relief cave is located on the right side of the table hole flood relief cave; the section is 6.5 m × 7.5 m in the shape of a city gate cave; the tunnel is 803.34 m long, the inlet floor elevation is 1,340 m; the maximum discharge capacity is 670.5 m³/s, using the pick flow energy dissipation. The power generation cavern adopts two machines and one cavern, a total of two single cavern water crossing capacities of 250 m³/s, and a single cited flow rate of 124.83 m³/s. The inlet adopts a shore tower type inlet, with one barrier, flat maintenance gate, and accident gate each. The orifice size is 8.0 m × 9.0 m; the inlet floor elevation is 1,350 m; the gate top platform elevation is 1,427 m; the hole diameter is 9 m; the hole length is 670.485 m and 660.995 m, respectively; the centerline spacing between the two holes is 32 m, the hole excavation volume is 113,800 m³.

The dam site area of Jilintai is a moderately acidic volcanic clastic rock of the Carboniferous system, mainly Yingan crystalline tuff, tuff breccia, andesite, etc. The thickness of the strong weathering layer of the rock is 3–5 m, and the depth along the fault is 10–20 m. The river valley in the dam site area is V-shaped, the slope of the topography on both banks is about 45°, the width of the valley at the normal water storage level is 340 m, the relative height difference between the tops of the two banks is 327–362 m, and the bedrock is mainly tuff. There are more than 700 large and small faults developed in the dam site area, with steeply dipping fissures and broken lithology. The largest F32 fault, which has a 20° angle with the river, extends upstream from the lower dam line to the right side of the riverbed and passes through the toe plate of the riverbed.

Jilintai-I Hydropower Station is located in an area with a high frequency of earthquakes and extremely cold weather, so the construction climate conditions. The project not only achieved high strength filling in the narrow river valley, but also ensured the anti-cracking and anti-freeze-thaw quality of the dam filling and toe plate panels, and through rigorous and scientific construction management and reliable technical measures, a series of challenging issues were solved, and the project was cut off ahead of schedule, accumulating richer experience for building such high dams in the western border of Xinjiang.

SUOFENGYING HYDROPOWER STATION

The First Batch of the "West-East Power Transmission Project" in Guizhou Province

Suofengying Hydropower Station is located in the middle reaches of the Wujiang River at the junction of Qianxi County and Xiuwen County in central Guizhou Province, 35.5 km upstream from Dongfeng Hydropower Station, 74.9 km downstream from Wujiangdu Hydropower Station, 54 km straight line distance from Guiyang City, Guibi High-Grade Highway passing through downstream 3.5 km from the dam site. It is located in the hinterland of the central part of Guizhou Province, the center of Guizhou's power load area, and is the second cascade of the planned tertiary power station on the main stream of Wujiang River. Suofengying Hydropower Station started one of the first "West-to-East Power Transmission" projects in Guizhou Province, and played a positive role in improving the utilization rate of water energy in Wujiang River Basin and relieving the power tension in Guizhou.

The pivot project consists of a compact concrete gravity dam, an open overflow meter hole, a right bank water diversion power generation system and underground plant, and other buildings, with a

Suofengying Hydropower Station panorama

control basin area of 2,186 km² above the dam site and a maximum dam height of 115.80 m. The normal reservoir storage level is 837 m, the dead water level is 822 m, the installed capacity is 600 MW, the guaranteed output is 166.9 MW, and the multi-year average power generation is 2.011 billion kW·h. The total reservoir capacity is 0.2012 billion m³, and the regulating reservoir capacity is 0.0674 billion m³, which is a daily regulating reservoir. The project is mainly for power generation, and also has the benefits of breeding and tourism. The power station is a large type II project. Its main building's dam, flood relief system, water diversion, and power generation system are 2-stage buildings, all designed for a 100-year return period flood; the dam and flood relief system are calibrated for a 1,000-year return period flood.

Suofengying Hydropower Station passed the feasibility study in September 2001, and the construction started in January 2001. The main construction preparations started in October 2001, the construction officially started on July 26, 2002, and the flow was cut off on December 18 of the same year. On December 28, 2003, the excavation for the main underground plant was completed, and the concrete for the plant foundation began to be poured. The first unit generated electricity in August 2005, the second unit in December of the same year, and the last unit was put into operation in early June 2006.

There are inter-river plots on the left and right banks of Wujiang Suofengying Hydropower

The Plateau Lake formed after the storage of water in Suofengying Hydropower Station

Station, which are composed of the main stream bay or tributaries, and the soluble rocks are widely distributed between the plots, and the overall direction is parallel to the river valley, with a gentle dip angle. The surface karst is developed, the regional fractures are crisscrossed, and the water barrier is mostly cut off, forming an unfavorable geological structure of soluble rocks distributed from inside the reservoir to the downstream outside the reservoir. The reservoir is composed of 2 sections of the main stream of Wujiang River and the tributaries of Cat Diving River. When the normal storage level is 837 m, the backwater length of the main stream section is 35.5 km, and the backwater length of the Cat Diving River section is 16 km, a narrow valley. It is a typical canyon-type reservoir.

As the reservoir of the project flooded about 104 km² of arable land and mountainous land, and 191 km² of land was requisitioned for construction and construction units, the project migration is the least among the Wujiang River basin, and very rare among hydropower projects in China. The project started the construction preparation in 2001 and achieved the power generation target at the end of 2005. As a large (II) hydropower station, it has a power generation period of 4 years and a total construction period of 5 years. The short construction period is extremely rare in the history of hydropower construction and is even comparable to the construction period of thermal power generation projects of the same scale. Due to the low inundation, low migration, reasonable hub arrangement, short construction period, and low investment, the project has become a very economically efficient water conservancy project.

By optimizing the design, the construction of the Wujiang cascade power station tried to avoid damage to the surface environment by large-scale open excavation. At the beginning of the design, the plant of Suofengying Hydropower Station was laid out on the ground on the left bank, and was adjusted to an underground plant on the right bank during construction, reducing the open excavation by nearly 600,000 m³ and protecting the vegetation on the nearly 300 m high slope. The plant has reduced 2.062 million m³ of excavated earth and rock, 720,000 m² of affected vegetation, and 300,000 m³ of excavation in the material yard, and reduced open excavation by more than 2 million m³. In addition, the power station has been greening the plant road and slopes by clearing dangerous rocks, spraying concrete slope protection, and planting vine plants. The greening of the gentle slopes is carried out with the three-dimensional greening of "trees + irrigation + grass," and the new ecological landscape is created according to the seasonal and color characteristics of plants. As for the slag and material yard after the construction, the slag removal, leveling, grass seed sowing, film laying, maintenance, replanting, etc., are adopted to plant suitable indigenous species to effectively restore the vegetation.

Suofengying Hydropower Station under construction

On June 6, 2006, Unit 3 was successfully put into operation, and all the hydropower stations were put into operation, marking the completion of the first batch of national "West-East Power Transmission" projects in Guizhou. The construction of Suofengying Hydropower Station overcame a large number of technical engineering problems and won many awards, such as "Report on Reservoir Karst Leakage of Wujiang Suofengying Hydropower Station" won the first prize of Guizhou Province Excellent Engineering Consulting Achievement in 2005, "Engineering Design of Suofengying Hydropower Station of Wujiang River in Guizhou Province" won the 15th First Prize of Guizhou Province Excellent Engineering Design in 2008, and "Engineering Geological Survey of Suofengying Hydropower Station of Wujiang River in Guizhou Province" won the 12th First Prize of Guizhou Province Excellent Engineering Survey in 2008. In addition, Suofengying Hydropower Station won the 10th China Civil Engineering Zhan Tianyou Award and the title of "2011 National Demonstration Project of Soil and Water Conservation in Production and Construction Projects," awarded by the Ministry of Water Resources.

PUBUGOU HYDROPOWER STATION
The World's Highest Gravel Soil Core Wall Rockfill Dam
on Deep Overburden

Pubugou hydropower station is located on the main stream of the Dadu River, the junction between Hanyuan County and Ganluo County, Ya'an City, in Sichuan Province. It is the national "Tenth Five-Year" key project and the landmark project of the grand western development program. It is also the largest single and total installed capacity hydropower station since its commissioning in the 21st century and during the Sichuan post-disaster reconstruction period. It is a controlled reservoir for the lower reaches of the Dadu River. Its main task is to generate electricity, with shared the role of flood control, sand control, and other comprehensive benefits of a mega hydroelectric hub project. The Pubugou dam is the world's highest gravel soil core wall rockfill dam built on the deep cover layer, with a maximum dam height of 186 m.

Pubugou Hydropower Station panorama

The project started construction in March 2004, and the first two units were put into operation at the end of 2009. The third unit was put into operation on April 7, 2010, the fourth unit on June 29, the fifth unit on December 23, and the sixth unit on December 26, 2010.

Pubugou Hydropower Station reservoir area

The Pubugou dam site has a narrow river valley, steep valley slopes, and complex geological conditions, which must be treated to meet the requirements for dam construction. The foundation of the dam is a deep cover layer with two vertical impermeable walls of 1.2 m each, the upstream wall is of the insertion type, and the downstream wall is of the gallery type connected with the core wall. The downstream wall is located on the dam axis, the center line of the two walls is 14 m apart, and the maximum depth of the impermeable wall below 670 m elevation is about 80 m. It is divided into four layers: drift pebble layer, pebble gravel layer, the sandy lens with drift pebble layer, and drift (block) pebble layer. The bedrock of the dam is divided by fault F2, with granite on the left bank and basalt on the right bank.

The Pubugou dam project solved a world problem of the wide-grade gravel soil for seepage control. The national scientific and technological research team made a breakthrough. By eliminating large particles to adjust the grade, increasing the soil compaction with a 25 t grade heavy bump mill, and strengthening the seepage outlet back filtration control and other supporting technologies, they successfully made innovative use of wide-grade gravel soil with less than 5% clay content for the core wall of the dam seepage control, a new type of seepage control materials. In view of the difficulty of building a high core wall rockfill dam on the deep cover layer, they proposed to use two large spacing high strength and low elasticity rigid concrete seepage control walls as the seepage control structure for the deep cover layer of the dam base. To reduce the impact of flood discharge atomization on the Chengkun Railway, they proposed a spillway hawksbill-type pick nosecone and overcame many technical problems. The project is of high quality, fast progress, and low investment, and has achieved the engineering goal of "safety, environmental protection, and energy saving," which has gained high recognition and social and economic benefits.

Pubugou hydropower station is regarded as the world's classic work of deep cover on the construction of a high heart wall rockfill dam. It solved many world problems in the dam heart wall impermeable material, dispersion of flooding mode, power station linkage water storage under the gate, and so on.

JIN'ANQIAO HYDROPOWER STATION
China's Largest Hydropower Station Invested in and Built by Private Enterprises

Jin'anqiao Hydropower Station is the fifth cascade of the "one reservoir and eight cascades" in the "Report on Hydropower Planning for the Middle Reach of Jinsha River" approved by the State Council, which is a national mega hydropower station with an annual generation capacity of about 13 billion kW·h after the completion of the leading upstream reservoir. It is one of the backbone power sources for the "West-East Power Transmission" and "Cloud Power Transmission to Guangdong." The project is jointly financed by Hanergy Holdings Limited, Yunnan Jinsha River Midstream Hydropower Development Company Limited, and Yunnan Development and Investment Company Limited, with a total static investment of 1.2501 billion yuan and a dynamic investment of 1.4679 billion yuan.

Jin'anqiao Hydropower Station is located in the middle reaches of the Jinsha River in Lijiang City, Yunnan Province, and is the fifth power station of the Jinsha River middle reaches "one reservoir and eight cascades" plan. Hub project is of large (I) first-class type, consisting of water retaining buildings,

Jin'anqiao Hydropower Station panorama

flood relief buildings, water diversion power generation buildings, and other components. Water retaining buildings are the crushed concrete gravity dam, the top of the dam elevation of 1,424 m, the maximum dam height of 160 m, the top of the dam length of 640 m, from left to right, there is the left bank non-overflow dam section, left bank sand flushing bottom hole dam section, plant dam section, right bank sand flushing bottom hole dam section, overflow table hole dam section, right bank non-overflow dam section. The project is mainly for power generation, with a normal reservoir storage level of 1,418 m, a corresponding reservoir capacity of 0.847 billion m³, an installed capacity of 2,400 MW, and the average multi-year power generation capacity of 11.043 billion kW·h.

In April 2002, the National Development Planning Commission approved the Jin'anqiao Hydropower Project. The project was launched in August 2003, and the pre-feasibility study report was approved on February 20, 2003. The river was cut off on January 9, 2006, and the power station officially generated electricity on March 27, 2011.

The river at the dam site of Jin'anqiao flows from north to south. It is straight, with a V-shaped longitudinal monoclinic valley. The topography on both sides of the river is basically symmetrical, with abundant mountains and steep terrain. The rock stratum in the dam site area is monoclinic, with hard basalt as the main structure. The bedrock in the riverbed area is mainly fractured chlorite, and the joint surface is filled with chlorite film, which forms rusty film after weathering, and the fractured chlorite is fractured and block-fractured. Groundwater in the dam site area is dominated by bedrock fracture diving, and the permeability of the rock body is uneven, with a general trend of gradual weakening from the surface to the depth.

The project was praised by the All-China Federation of Industry and Commerce as a landmark project for private enterprises to enter the field of large-scale national infrastructure construction, and is the largest and only megawatt hydropower station of installed capacity built and operated by private enterprises so far. As a model for private enterprises to build large-scale national projects, the scale of which was unprecedented. Because of its location in a deep mountain

Jin'anqiao Hydropower Station flood release

valley, submerged arable land, and the small number of immigrants, Jin'anqiao Hydropower Station was the lowest comprehensive cost indicator for the construction of a power station on the Jinsha River at that time, and was also the first large hydropower station on the main stream of the Jinsha River to be connected to the grid for power generation.

GANGNAN HYDROPOWER STATION

China's First Hybrid Pumped Storage Power Station

Gangnan Hydropower Station, located in the middle reaches of the Hutuo River, Gangnan Village West, Ping Shan County, is a large water conservancy project to manage the Hutuo River, regulate floods, and develop and utilize water resources. The main role of the project is flood control and irrigation, taking into account power generation, water supply, and aquaculture. The dam site controls a basin area of $15,900$ km², accounting for $2/3$ of the Hutuo River mountainous area. The total capacity of the reservoir is 1.571 billion m³.

On March 10, 1958, the project started construction, and on July 15, 1959, the top of the main and sub dams were filled to 197 m elevation, and for the first time, played a flood storage benefit. In 1962, according to the national adjustment policy, the construction was stopped and transferred to maintenance, and the two 15 MW units started to operate for power generation.

In October 1966, the reservoir renewal project was started. The renewal project includes the main and sub dams raising; normal spillway reconstruction; the remaining concrete pouring of the extraordinary spillway and the renewal of the traffic bridge; the new flood relief cave construction, with a diameter of

Gangnan Hydropower Station panorama

5.4–6 m; and the installation of an energy storage generator set. By the end of 1969, the renewal project was basically completed as scheduled.

The main buildings of the project include the main dam, sub dam, water transfer cavern, flood relief cavern, normal spillway, extraordinary spillway, regulating pond, additional spillway, etc.

(1) Main dam. It is a clay-inclined wall dam with a crest elevation of 209 m, a maximum dam height of 63 m, and a dam length of 1,701 m.

(2) 17 sub dams. Twelve on the left bank and five on the right bank, with a total length of 4,757 m, all of which are homogeneous earth dams.

(3) Water transmission cave. It is a flood relief and power generation tunnel with a diameter of 6 m. The power generation hole is 389 m long, with three forks at the end and three generating units installed with a total capacity of 41 MW. The flood relief branch tunnel is 155.8 m long, with a dissipation pool at the outlet. 5 m × 6 m flat steel gates are installed at the inlet, and 4.5 m × 4.5 m arc gates at the outlet of the water transmission hole. The maximum water flow in the flood relief branch cave is 388 m³/s when the gate is opened at 3.5 m.

(4) Flood relief cave. It is 698.4 m long, of which the diameter of the hole is 5.4 m in the 155.4 m long front section. In the back section, the diameter of the pressurized hole section is 6 m, and a 6 m × 6.5 m door-shaped gate is in the section of the non-pressurized hole. The tail cave is connected with the pressureless cave by the flood relief nullah, without energy dissipation at the outlet. The inlet is installed with 4.5 m × 6 m flat steel gates, and the outlet is installed with 4.5 m × 4.5 m arc gates, with a maximum flood flow of 468 m³/s. A hole is left in front of the outlet gates for the water transfer to the diversion canal.

(5) Normal spillway. The normal spillway is open with gates control. It has four holes, each net width of 12 m per hole, with four arc gates of 12 m × 12.3 m. The maximum flood discharge capacity is 5,640 m³/s.

(6) Emergency spillway. An emergency spillway is open for the riverbank with a net width of 41.6 m, without gates installed. The top elevation of the weir is 194 m, and a clay-inclined wall-type earth dam is built in front of the weir, the top elevation of which is 205 m. In the event of a once-100-year flood, the dam will release flood water with a maximum discharge of 3,520 m³/s.

(7) Regulating pool. Used for hydroelectric power station storage power generation and regulation of water from the Wentang River, water supply to Dachuan Canal, Beiyue Canal, Shuilun pump station, and Baji station.

(8) New spillway. Occupying the left bank No. 1 vice location, it was built as an open type on the river bank after the flood in 1978. It has eight holes, each with a net width of 9 m, installed with 9 m × 15.5 m arc gate control, and the top elevation of the weir was 191 m. In September 1986, the project above the bucket and the supporting management house and backup power supply facilities were completed. The second phase of the project was completed in June 1989. This project was designed for a 1,000-year return period flood and calibrated for a 10,000-year return period flood.

FENGNING PUMPED STORAGE POWER STATION

The Pumped Storage Power Station with the World's Largest Installed Capacity

Fengning Pumped Storage Power Station is located in Fengning Manchu Autonomous Region, Chengde City, Hebei Province, 180 km south of Beijing and 170 km southeast of Chengde City. After its completion, it will solve the problem of insufficient peaking regulation capacity of the Beijing-Tianjin and northern Hebei power grids, working jointly with the previously built pumped storage power stations like that in Ming Tomb Reservoir and other peak regulating power supplies. At the same time, according to the demand of the power grid, the power station can also undertake the system frequency regulation, phase regulation, load backup, and emergency backup tasks, maintaining the safe and stable operation of the power grid.

Fengning Pumped Storage Power Station under construction

Fengning Pumped Storage Power Station has a total installed capacity of 3,600 MW and is developed in two phases, with the installed capacity of both the first and second phase projects being 1,800 MW. The pivot building mainly consists of the upper and lower reservoirs, the water transmission system of the first and second phase projects, and the underground plant and switching station.

This project was proposed in 1995 and formally approved on August 21, 2012, by the National Development and Reform Commission after 17 years of unremitting efforts by the parties involved. The first phase of groundbreaking was in May 2013. On September 23, 2015, the mobilization meeting for the start of the Hebei Fengning (Phase II) pumped storage power plant project was held in Beijing.

The runoff from the basin above the lower reservoir site of Hebei Fengning Pumped Storage Power Station is mainly recharged by rain and snow melt water, and there are two abundant water periods every year. The first period of abundant water is in April, with snow and ice melt water as the main source of water. The second is from June to October, and the main source is rainfall. Since most of the watershed above the dam site belongs to the grassland area above the dam, the terrain is flat, which has a large effect on runoff storage, making an insignificant inter-annual variation of runoff. According to the natural runoff series of the restored Fengning dam site, the multi-year average flow is 7.60 m³/s, the multi-year maximum flow is 59.3 m³/s (August 1959), and the multi-year minimum flow is 0.341 m³/s (January 2007). The maximum annual average flow is 18.6 m³/s (1959–1960), and the minimum annual average flow is 4.68 m³/s (1989–1990), with the maximum annual average flow being about four times the minimum annual average flow. The annual distribution of runoff is most abundant in April, accounting for 16.3% of the year, followed by July and August, 15.0% of the year in August during the main flood season, 58.0% of the year in June to October during the flood season, and 42.0% of the year from November to May the next year.

The power station is connected to the Beijing-Tianjin-Hebei northern power grid through two 500 kV lines, which can provide large-capacity energy storage services for the system, promote the large-scale development of wind, solar and other renewable energy sources in northern Hebei, and play an important role in meeting the peak regulation demand of the power grid, improving the quality of power supply and the economy of power grid operation, and promoting the effective use of resources. At the same time, the power station can also effectively improve the power supply structure of the grid, enhance the system's peaking capacity, and promote the development and power consumption of Hebei's 10-megawatt wind power base. The project of Fengning pumped storage power station boosts green development and benefits the people, boosting local GDP by 30 to 40 billion yuan.

China's pumped storage power plant construction has been at the forefront of the world. The current world's largest pumped storage power plant under construction is the Fengning Pumped Storage Power Station in Hebei, with an installed capacity of 3.6 million kW, equivalent to about 1/6 that of the Three Gorges Power Station. The power station plans to put the first unit into operation in 2022. As a national key project, Fengning pumped storage power plant is also known as the world's largest "super-charger," which can consume 8.8 billion kW·h of excess electricity per year, with an annual power generation capacity of 6.612 billion kW·h. It can save 4.88 million t of standard coal and reduce 1.2 million t of carbon emission, which is equivalent to the reforestation of more than 240,000 mu.

YIXING PUMPED STORAGE POWER STATION

The First Pure Pumped Storage Power Station with an Installed Capacity of 1 Million kW in Jiangsu Province

Yixing Pumped Storage Power Station is one of the important energy projects built during the "Tenth Five-Year Plan" period in Jiangsu Province and the first daily regulating pure pumped storage power station with an installed capacity of one million kilowatts. The power station is located in the scenic Tongguan Mountain area in the southwest suburbs of Yixing City, Jiangsu Province, about 10 km away from Yixing City, with an installed capacity of 1 million kW, and is mainly responsible for peak regulation, valley filling, frequency regulation, phase regulation and rotary standby in the power grid.

Yixing Pumped Storage Power Station upper reservoir panorama

The power station consists of an upper reservoir, lower reservoir, underground plant, water transmission system and ground 500 kV switching station, and other buildings. The upper reservoir is built on the top of Tongguan Mountain, with a total storage capacity of 5.307 million m³, an effective storage capacity of 5.073 million m³, a normal storage level of 471 m, and a dead water level of 428.6 m. It is a reinforced concrete panel with full reservoir basin impermeability. The main dam is a concrete panel rockfill dam with a crest elevation of 474.2 m, a crest length of 495 m, and a maximum height of 75 m; the sub dam is a crushed concrete gravity dam with a maximum height of 34.9 m, a crest length of 220 m and a crest elevation of 474.2 m.

The lower reservoir is located in the northeast foothills of Tongguan Mountain, with a total storage capacity of 5.728 million m³ and an effective storage capacity of 5.223 million m³; the water retaining building is a clay core wall rockfill dam, with a top elevation of 83.4 m, a top length of 483 m and a maximum dam height of 50.4 m; the normal storage level is 78.90 m, and the dead water level is 57 m. The underground plant cavity group is located in the middle of the water transmission system, with a burial depth of about 320 m, and the excavation size of the main plant is 155.3 m × 22.0 m × 52.4 m (length × width × height). The water transmission system is set up in the mountain between the upper and lower reservoirs, with two single-length 2,800 m diversion tunnels, using two holes and four machines; the tailwater system uses two machines and one hole form. The ground switch station is located on a relatively gentle hillside above the underground plant cavern.

The power station is equipped with four single-stage plasticized mixed-flow pump turbine generating motors with a capacity of 250,000 kW. When the power grid is in the low season, it can pump water from the lower reservoir to the upper reservoir with a pumping capacity of 1 million kW, with an annual pumping power consumption of about 1.96 billion kW; when the power grid is in the peak season, it takes only three or four minutes to draw water from the upper reservoir to generate electricity, with an annual power consumption of 1.49 billion kW·h. The power station is connected to the 500 kV Yixing substation by two 500 kV transmission lines, which directly feed into the main grid of East China and Jiangsu, supplying power to the Jiangsu power grid. The power station optimizes the power structure of the power grid, enhances the ability of peak regulation and valley filling, ensures the stable, safe, and economic operation of the power grid, and strongly promotes the continuous, stable, and rapid development of the economy.

The main project of the Yixing Pumped Storage Power Station started in August 2003, and all four units were put into operation in 2008, setting a new pumped storage construction record of putting all four units into operation in one year, 13 months earlier than the state-approved construction period. The design and construction quality of the clay core wall rockfill dam of the power station reached the advanced international level. It is the first time in China to independently develop and successfully press a variety of types of copper water stop joint molds as a whole. In addition, six construction methods were rated at the national level: the concrete panel pile rock dam filling, rock wall crane beam concrete construction, concrete panel pile rock dam panel construction, crushed concrete silo surface construction, crushed concrete dam construction in the metamorphosis of concrete construction, rock wall crane beam rock platform (two-way controlled explosion method) excavation construction. After its completion, the power station has become a supporting power source for the load center in southern Jiangsu, which can replace coal power units to undertake the task of rotating backup and emergency backup of the power grid and save the no-load coal consumption of coal power units.

LIYANG PUMPED STORAGE POWER STATION
The Largest Pumped Storage Power Station in Jiangsu Province

Liyang Pumped Storage Power Station is located in Liyang City, Jiangsu Province, which is a large first-class project and is the largest installed pumped storage power station in Jiangsu Province. The main task of the power station is to provide peak regulation, valley filling, and emergency accident backup for the power system of Jiangsu Province, and at the same time, it can undertake the task of frequency regulation and equalization of the system. Liyang pumped storage power station exploration started in 2002, construction started in December 2008, and it was completed and put into operation in October 2017 after nine years of construction. In June 2020, the Acceptance Committee of the General Institute of Water Resources and Hydropower Planning and Design agreed to pass the project completion inspection and acceptance.

The upper reservoir of Liyang Pumped Storage Power Station panorama

The total investment of the Liyang Pumped Storage Power Station is 8.92 billion RMB. The pivot building of the power station consists of four parts: upper reservoir, lower reservoir, water transmission system, and powerhouse system. Facing the difficulties and challenges such as complicated geological conditions, construction difficulties, and long construction periods, all the participants carry forward the spirit of "building first-class projects by unity and cooperation," putting safety, quality, and progress as the top priorities to ensure orderly progress of the project.

The upper reservoir is located in the Wuyuanshan work area of the Longtan forestry farm. The overall trend of its topography is high in the west and low in the east, the west bordering Langxi County, Anhui Province. The normal storage level of the reservoir is 291 m, the dead water level is 254 m, the total storage capacity is 14.23 million m³, and the regulating storage capacity is 11.95 million m³. The main and sub dams of the upper reservoir are reinforced concrete panel rockfill dams. The top elevation of the dam is 295 m, the maximum height of the main dam is 165 m, and the total length of the top of the dam is 1,113.198 m.

Liyang Pumped Storage Power Station upper reservoir

The lower reservoir is located in Wu Village, Tianmu Lake Town, adjacent to Zhongtianshe River, a tributary of the south source of Shahe Reservoir, excavated in the river mound terraces, wide and shallow gullies, and residual mounds, and dammed on the side of the adjacent Shahe Reservoir, where the excavated material is mainly used for damming the upper reservoir. The lower reservoir is L-shaped, with a normal storage level of 19 m and a dead water level of 0 m, with a total storage capacity of 13,440 m³. The lower reservoir water retaining building is a homogeneous earth dam, with a maximum height of 12.60 m, and the water replenishment and discharge gate is located at the right head of the dam. The width of the orifice is 4 m, and two flat gates are set to control it. A total of 18 million m³ of rock is excavated in the lower reservoir, which is equivalent to the volume of 900,000 trucks with 50 t load.

The water transmission system is arranged downstream of the left head of the main dam of the upper reservoir, and the diversion and tailwater both adopt the joint water supply method of one hole and three machines. The import and export of the upper and lower reservoirs adopt the vertical shaft type and the shoring tower type, respectively. The main hole diameter of the two diversion tunnels is 9.20 m, and the whole section is lined with steel plates; the main hole diameter of the two tailwater tunnels is 10 m, and they are lined with reinforced concrete. The tailwater regulating chamber is a circular impedance-type regulating chamber with a diameter of 22 m.

The underground plant adopts the head-type development method, and the vertical burial depth of the main plant is 240–290 m. The underground cavern group mainly consists of the main plant, the main transformer cave, the busbar cave, the cable shaft of the high-voltage cable machine, etc. The excavation size of the main plant is 291.90 m × 23.50 m × 55.30 m (length × width × height), and the excavation size of the main transformer cave is 193.16 m × 19.70 m × 22 m. The switch station is located in a relatively gentle section downstream of the main plant, with a ground elevation of 88.90 m and a plane size of 123 m × 82 m (length × width). The power station is equipped with six reversible pump hydro generator sets with a single capacity of 250 MW, designed for annual power generation of 2.007 billion kW·h and annual pumping power of 2.676 billion kW·h. The rated head of the power station is 255 m, and the cave length-to-head ratio is 7.9.

Liyang Pumped Storage Power Station lower reservoir

The construction of the Liyang Pumped Storage Power Station has many highlights worthy of praise. The six units with 250 MW capacity and 1,500 MW total capacity rank first in Jiangsu Province, and the six units were put into operation in the same year, creating a record of "six units put into operation continuously in one year." The reversible pump turbine adopted by the power station is a localized unit designed, manufactured, and supplied by the domestic host manufacturers independently, which makes Liyang Pumped Storage Power Station the first power station in Jiangsu Province to adopt a localized unit. The design and manufacturing difficulty of the unit are high, and the maximum operating head variation of the power station reaches 1.29, which is the largest pumped storage unit with the largest head variation put into operation in China. The vertical shaft inlet and outlet of the upper reservoir is the first in Asia

Upper reservoir shaft type inlet and outlet

and rare in the world in terms of construction volume and difficulty. Compared with the coal-fired or gas turbines and other kinds of power sources built in the grid, Liyang Power Station is an ideal peaking power source for the Jiangsu power grid, which is equivalent to building a "large battery" for the Jiangsu power grid. It can not only optimize the power supply structure, improve the power supply quality and reliability of the grid, but also save the system power supply construction funds and operating costs, which can positively promote the reduction of grid coal consumption, energy saving, and emission reduction, and green development. As the largest pumped storage power station in Jiangsu Province, it has reduced about 400,000 t of CO_2 per year since its operation, and thus its effect on ecological protection is obvious.

YAMDROK LAKE PUMPED STORAGE POWER STATION

The World's Highest Altitude Pumped Storage Power Station

Yamdrok Lake Pumped Storage Power Station is located in Gonga County, Tibet Autonomous Region, 80 km from Lhasa City. It is a hybrid pumped storage power station mainly for supplying power to Lhasa, Shannan, and Shigatse. It is also responsible for peak regulation, frequency regulation, and accident backup of the Lhasa power grid. The ground elevation of the power plant is about 3,600 m, which is the highest altitude pumped storage power station in the world and the highest head-pumped storage power plant in China. The power station uses the natural drop of more than 840 m between Yamdrok Lake and Yarlung Tsangpo River to take the water from Yamdrok Lake and divert it to the power plant by Yarlung Tsangpo River through the water diversion tunnel and pressure steel pipe.

Yamdrok Lake Pumped Storage Power Station

Yamdrok Lake Pumped Storage Power Station is a key project of the national "Eighth Five-Year Plan," which was started in September 1989, and the first unit was connected to the grid on June 1, 1997, and completed in December of the same year.

The main buildings of the power station are a water inlet by the Yamdrok Lake, a water diversion tunnel, a pressure regulating well, a pressure pipeline, a ground-type plant and 110 kV switching station, and water intake by the Yarlung Tsangpo River, a sand sink, a pumping steel pipe connected with multi-stage energy storage pump. When the hub is generating electricity, water is taken directly from Yamdrok Lake and discharged into the Yarlung Tsangpo River through tunnels, pressure regulating wells, and pressure pipes to the tailwater of the plant; when pumping is in operation, water is pumped from the riverside low-head pump house into the sand sink and then into the multi-stage energy storage pump of the main plant, flowing into Yamdrok Lake through the water diversion system.

The upper reservoir is the largest sacred lake in Tibet—Yamdrok Lake, with a capacity of 15 billion m³ and abundant incoming water. The water level variation is only 1.23 m during the year, which can guarantee the water level in the lake and the original ecological environment of the lake area. The lower reservoir is actually the Yarlung Tsangpo River. During the operation of the power station, generally speaking, the water of Yamdrok Lake is not used, and the water of the lower reservoir by the Yarlung Tsangpo River can be pumped to Yamdrok Lake for power generation in the dry season. The total installed capacity of the power station is 112.5 MW, with an average annual power generation capacity of 91.8 million kW·h and annual power generation utilization hours of 1,000 h.

The water diversion tunnel is a circular pressurized tunnel, wholly lined with reinforced concrete. The pressure steel pipe adopts one main pipe for water supply, the front section is a buried pipe with a length of 754.4 m, and the back section is an open pipe with a length of 2,290.4 m. The high-pressure section in the open pipe is made of a high-strength steel plate with a tensile strength of 610 MPa imported from Japan. The main plant is 69.8 m long, with four sets of three-machine pumped storage units made by ELIN-VOITH (Austria), and one conventional unit is reserved for the location. The sand sink is located by the Yarlung Tsangpo River, through which the sediment particles with a particle size of more than 0.1 mm in the water of Yarlung Tsangpo River can be precipitated by more than 80%, thus reducing the wear and tear of sediment on the units.

Since its operation, Yamdrok Lake Pumped Storage Power Station has made a great contribution to the peaking regulation and valley filling of the Lhasa power grid and the emergency standby. Although the investment per unit of power is high, the operating cost is lower than that of other power stations. It is still an economic power source in the Tibetan region. In addition, over the years of operation, the environmental changes in the Yamdrok Lake basin have been mainly caused by natural factors. The scope and degree of changes influenced by humans and engineering activities are relatively small.

GUANGZHOU PUMPED STORAGE POWER STATION

China's First Pumped Storage Power Station

Guangzhou Pumped Storage Power Station is located in Conghua District, northeast of Guangzhou City, about 90 km away from Guangzhou City. The power station was built in two phases to enable the smooth and safe operation of the Daya Bay Nuclear Power Station in Shenzhen and to undertake the task of peak regulation, frequency regulation, phase regulation, and accident backup of the Guangzhou power grid.

Guangzhou Pumped Storage Power Station upper reservoir panorama

The construction of Phase I of the Guangzhou Pumped Storage Power Plant started in September 1988, and the main project was formally started in May 1989. The second phase of the project was officially started in September 1994, the first unit generated electricity in April 1999, and all four units were put into commercial operation in June 2000.

The main buildings of Phase I and Phase II projects are an upper reservoir, lower reservoir, water diversion system, power plant, and 500 kV switching station. The first and second phases of the project share the upper and lower reservoirs, and the distance between the two reservoirs is 4.2 km. Both the upper and lower reservoirs are supplemented by natural runoff, and the Zhaoda Water of the upper reservoir and the Jiuqu Water of the lower reservoir are both tributaries of the Niulan River in the upper reaches of the Liuxi River.

The first and second phases of the project are equipped with four reversible pump turbines with a single capacity of 300 MW (power generation condition) and a total installed capacity of 2,400 MW, which is the largest pumped storage power station in the world in terms of installed capacity.

Guangzhou Pumped Storage Power Station successfully adopted advanced technology during the construction process, achieving the goals of a short construction period, good quality, and saving investment. The second phase of the project is based on the experience of the first phase and some optimization of design and construction. For instance, the tailwater regulating wells of the diversion system for the first phase is in the form of two machines and one well, a total of two wells. In the second phase it was changed to four machines and one well, and the operation practice proved to be effective. The high-pressure inclined shaft was lined with slip-form construction, setting a new construction record of 207 m monthly. The light support parameters of the underground powerhouse and large cavern have reached the advanced international level. The underground plant structure adopts thick slab beams below the generator level and a monolithic wall system embedded in the surrounding rock, which improves the dynamic characteristics of the structural system in absorbing unit vibration and improving the operating conditions, reflecting the progress of contemporary pumped storage power plant technology.

Since its operation, it has exerted remarkable benefits in the Guangzhou power grid. Take Phase I power plant as an example; it absorbs 1.405 billion kW·h of low valley power and 1.08 billion kW·h of peak power generation annually on average, which can serve in peak regulation and valley filling, frequency and phase regulation for the power grid, with an average annual running time of 2,217 h per unit and 2.25 starts per unit per day on average. When the system has an accidental perimeter wave below 49.8 Hz, the average annual emergency start-up is 16.5 times. In addition, the reliability of the unit is also high. In 1999, the success rate of generation start-ups reached 99.8%, and pumping start-ups reached 97.7%. It only takes 2 mins from standstill to full power generation and 4 mins from standstill to full pumping capacity for the second phase unit.

TIANHUANGPING PUMPED STORAGE POWER STATION

The Power Station with the Highest Water Drop in the World

Tianhuangping Pumped Storage Power Station is located in Anji County, Zhejiang Province, on Daxi, a tributary of the West Tiaoxi River in the Taihu Lake Basin, 57 km from Hangzhou City, and is responsible for peak regulation, valley filling, frequency regulation, phase regulation and accident backup of the East China power grid.

The power station hub includes the upper reservoir, lower reservoir, water transmission system, underground plant cavern group, and switching station.

The total installed capacity of the power plant is 1,800 MW. The preparatory works started in June 1992, and the main construction started on March 1, 1994. Unit 1 was put into trial operation on September 30, 1998, and the last unit was put into operation in 2000.

Tianhuangping Pumped Storage Power Station upper reservoir panorama

The upper reservoir is dug and filled by a natural depression. The design maximum storage level is 905.2 m, and the corresponding capacity is 9.192 million m³. The upper reservoir consists of one main dam and four sub dams. All are earth and stone dams. The maximum height of the main dam is 72 m. The main and sub dams and the bottom seepage prevention use asphalt concrete, and the total area is 286,800 m².

The basin area above the dam site is 24.2 km², with an average annual runoff of 27.6 million m³. The designed maximum storage level is 344.5 m, with a corresponding capacity of 8.5956 million m³. The dam is designed for a 100-year flood and calibrated for a 1,000-year return period flood, and the maximum possible flood does not roam the top. The side weir spillway without gate control is set in the left head of the dam.

The water transmission system is excavated in the rhyolitic felsic tuff rock body. The upstream water transmission system is arranged with two 7 m inner diameter inclined shaft-type high-pressure tunnels lined with reinforced concrete. They are connected to three steel-lined branch pipes through reinforced concrete fork pipes to divert water into the pump turbines, the maximum hydrostatic pressure of the water transmission system is 680.2 m head, and the maximum dynamic water pressure reaches 887 m head. Each unit is connected to a tailwater tunnel, with a steel plate lining the downstream gradient section of the tailgate chamber.

The 500 kV switch station platform is arranged at an elevation of 350.2 m above the water inlet (outlet) on the left bank of the lower reservoir. The main and sub plants, main transformer room, busbar cavern, 550 kV cable shaft, tailgate room and traffic, ventilation, drainage, and other cavern groups are arranged in the rhyolitic fused tuff rock body. All the underground caverns are supported by spray anchors and locally reinforced steel mesh. Six 300 MW units are installed in the plant, with an annual power generation capacity of 3.014 billion kW·h and an annual pumping capacity of 4.104 billion kW·h. The turbine is a vertical shaft single-stage reversible mixed-flow pump turbine, with a rated output of 306 MW when the rated net head of the turbine is 526 m, and the maximum input power of the pump is 336 MW. The power generation motor is a vertical shaft suspended, air-cooled reversible three-phase synchronous motor, with rated capacities of 333 MW and 336 MW for generator and motor conditions, respectively.

Tianhuangping Pumped Storage Power Station has the highest head of a single-stage pump turbine head among all the existing pumped storage power stations built in China. It also ranks among the top similar power stations in the world. The upper reservoir basin adopts asphalt concrete anti-seepage protection, and 500 kV cable adopts dry cable, which is the first case in China at that time.

JIANGXIA TIDAL POWER STATION

China's First Two-Way Tidal Power Station

Jiangxia Tidal Power Station is located in Jiangxia Port, the top branch of Yueqing Bay, Zhejiang Province, in Wenling City. The power station was rebuilt by the Qiyitang Reclamation project, which had not yet been completed at that time. It is an experimental two-way tidal power station built in China in the 1980s, which is also China's largest tidal power station. The power station project also has comprehensive benefits such as reclamation, breeding, and transportation. The power station started construction in October 1972, the first unit generated electricity in May 1980, and the construction was fully completed in December 1985.

Jiangxia Tidal Power Station panorama

Jiangxia Port, where the power station is located, is 9 km deep and 686 m wide at the entrance of the dam site. It is a long, narrow, semi-closed shallow, semi-daily tide port. It is located in China's high tide difference zone; the multi-year average tide difference is 5.08 m; the maximum tide difference is 8.39 m; the minimum tide difference is 1.53 m. The reservoir capacity below the normal storage level is 5.14 million m³, and the effective reservoir capacity for power generation is 3.36 million m³. The power station adopts the single reservoir two-way mode, and the hub building is composed of a dam, sluice, plant, switching station, etc.

The dike is built on a 46 m thick layer of saturated marine silt clay with a clay core wall rockfill dam with a maximum height of 15.5 m. The sluice gate is located between the dike and the plant and is a 5-hole flat bottom sluice gate built on tuff, reinforced by the original Qiyitang Reclamation project drainage gate according to the bi-directional operation of the power plant. The power plant house is built on the left bank of the sluice gate, avoiding the complexity of foundation treatment set on the soft foundation of the seabed. It is equipped with five units, with a total capacity of 3.2 MW. Another pit is reserved for testing new tidal power generation technology, which can be expanded to 3.9 MW. The power station adopts a tubular bulb unit, which can generate power and discharge water in both directions. The bi-directional operating conditions can be realized without shutdown by adjusting the rotating angle of the paddles and the opening degree of the guide vanes.

Jiangxia Power Station is a scientific experimental power station, which mainly researches the characteristics of tidal energy, metal problems of offshore buildings, the development of tidal generator sets, and comprehensive utilization. The construction and operation of the power station have achieved certain scientific research results and gradually accumulated experience. Due to the measures of applied current cathodic protection and anti-adhesion coating, there is basically no corrosion of metal components and sea life adhesion in the seawater environment, and no abnormal phenomenon occurs to the unit's overall structure. After all five units were put into operation, the annual power generation reached a maximum of 6.46 million kW·h and a minimum of 5.04 million kW·h, with a multi-year average of 5.7 million kW·h. Due to the influence of various factors, the maintenance time was long, and the average annual utilization hours of the installed units were only 1,781 h.

YANGTZE RIVER THREE GORGES WATER CONSERVANCY HUB PROJECT

The Largest Hydraulic Engineering Project in the World Today

Yangtze River Three Gorges Water Conservancy Hub Project ("Three Gorges Project") is located in Sandouping, Yichang City, Hubei Province, in the Xiling Gorge, the main stream of Yangtze River, 38 km away from the Three Gorges Exit Nanjin Pass, and 40 km downstream of the Gezhouba Water Conservancy Hub. The project is the largest water conservancy project in the world today, with comprehensive benefits of flood control, power generation, and shipping.

Three Gorges Hydropower Station panorama

The main buildings of the hub include a dam, hydroelectric power station, flood relief buildings, and navigation buildings (five-stage locks and boat lifts). The dam site controls a watershed area of 1 million km² and a multi-year average flow of 14,300 m³/s. Total reservoir capacity of 39.3 billion m³, including flood control reservoir capacity of 22.15 billion m³. The design flood standard for the major buildings in the hub is a 1,000-year return period, flood flow of 98,800 m³/s; calibration flood standard for a 10,000-year return period flood plus 10%, peak flood discharge of 124,300 m³/s.

China has a long history of envisioning and exploring the construction of the Three Gorges Hydraulic Hub Project. In the early 20th century, Dr. Sun Yat-sen proposed the idea of developing the Three Gorges hydraulic resources, and in 1944, the China Resources Committee, in collaboration with Savanche of the U.S. Bureau of Reclamation and others, conducted a study of dam construction options and presented a preliminary report on the Three Gorges plan for the Yangtse River with a dam at Nanjin Pass. After the founding of the People's Republic of China, the preliminary work of the Three Gorges Project was carried out, and the Yangtze River Water Resources Committee of the Ministry of Water Resources did a lot of surveying, scientific research, planning, and design work. In 1986, the former Ministry of Water Resources and Electricity organized experts from various fields to discuss the feasibility of the Three Gorges Project, and concluded that the Three Gorges Project could play an irreplaceable role in flood control in the middle reaches of the Yangtze River, bring enormous benefits of power generation and shipping, and solve immigration and environmental problems. The Three Gorges Project should be built as soon as possible. Based on the results, the Yangtze River Water Resources Committee of the Ministry of Water Resources proposed a feasibility study for the Three Gorges Project in 1989, which was reviewed and approved by the State Council at the Fifth Session of the Seventh National People's Congress on April 3, 1992, and included the construction of the Yangtze River Three Gorges Water Conservancy Hub in the ten-year plan for national economic and social development. After a comparative study of various options, it was decided to adopt the implementation plan of "first-level development, one-time construction, phased water storage and continuous migration."

The Three Gorges Project is planned to last 17 years and will be completed in three phases. The first phase (1993–1997) is the construction preparation and the first phase of the project, marked by the interception of the Grand River; the second phase (1998–2003) is the second phase of the project, marked by the initial reservoir storage, the first units of power generation and the opening of the permanent locks; the third phase (2004–2009) is the third phase of the project, marked by the completion of all units of power generation and the construction of the hub. On May 20, 2006, the last concrete bin was poured, marking the completion of the entire Three Gorges Dam.

The dam is a concrete gravity dam with a crest elevation of 185 m and a maximum height of 175 m. The sub dam on the right bank of Maoping Creek is an asphalt concrete core wall gravel dam with a maximum height of 104 m. The total installed capacity of the hydropower plant is 18,200 MW. In addition, the underground plant is reserved on the right bank of the mountain, and six units with a capacity of 4,200 MW are to be expanded later.

The navigable buildings include permanent locks and a ship lift. The permanent lock is a double-line five-level lock, which can pass through the 10,000-ton fleet; the ship hoist is a single-line first-class vertical lifting type, which can pass through 3,000 t-class passenger and cargo ships.

During the construction, the builders overcame many world-class technical problems, such as the interception of the Grand River, the interception of the open diversion channel, the construction of deep-water cofferdams, the excavation of the upright high slope of the locks, and the control of stable deformation, the concrete pouring, the installation of extra-large metal structures and extra-large hydro generator sets, etc. As the world's largest hydraulic hub project, many of the design indicators of the Three Gorges Project have broken the world record for hydraulic engineering.

(1) The total reservoir capacity of 39.3 billion m³, with a flood control capacity of 221.5 billion m³, reservoir flood regulation can reduce the peak flow of 2.7–3.3 million m³/s. It is the water conservancy project with the world's most significant flood control benefits.

(2) It is the world's largest hydropower plant, with a total installed capacity of 18,200 MW and an average annual power generation capacity of 84.68 billion kW·h.

(3) It is the world's largest water conservancy project in terms of both individual and overall construction scale, with the total length of the dam axis of 2,309.47 m, the length of the discharge dam section of 483 m, and the two-lane 5-stage locks + boat lift.

(4) It is a water conservancy project with the world's largest volume of construction. The volume of earth and rock excavation and filling for the main building of the project is about 0.134 billion m³, the volume of concrete pouring is 27.94 million m³, 463,000 t of steel fabrication and installation, and 256,500 t of metal structure fabrication and installation.

(5) In 2000, the amount of concrete poured was 5.4817 million m³, with the highest monthly pouring volume reaching 550,000 m³, setting a world record for concrete pouring and making it the most difficult water conservancy project in the world to be constructed.

(6) It is a water conservancy project with the largest flow during the construction period, with an interception flow of 9,010 m³/s, and construction diversion of the maximum flood flow of 79,000 m³/s.

(7) The floodgate is the world's largest, with a maximum flood discharge capacity of 102,500 m³/s.

(8) The two-line, five-stage locks of the Three Gorges Project, with a total head of 113 m, makes it the inland locks with the largest number of stages and the highest total head in the world.

(9) The effective size of the ship hoist is 120 m × 18 m × 3.5 m, the maximum lift is 113 m, the weight of the shipping box with water reaches 11,800 t, and the overboard tonnage is 3,000 t, which is the largest and most difficult ship hoist in the world.

(10) The dynamic migration of reservoirs can eventually reach 1.13 million people, which is the world's largest number of reservoir migrants, making it the most arduous immigration construction project.

The Three Gorges Project has significant flood control, power generation, and shipping benefits. In terms of flood control, it can effectively control floods in the upper reaches of the Yangtze River, and has significant benefits for flood control in the middle and lower reaches of the plain, especially in the Jingjiang River area, raising the standard of flood control in the Jingjiang River section from a 10-year return period to a 100-year return period.

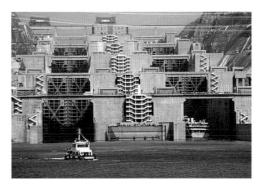

Three Gorges Project double-line five-level locks

In case of a 1,000-year return period flood or larger flood, cooperating with the Jingjiang River flood diversion projects can prevent the devastating disaster of the Jingjiang River embankment break; it can also greatly improve the mobility and reliability of flood control dispatching in the middle and lower reaches of the Yangtze River, and reduce flood inundation losses in the middle and lower reaches and the flood threat to Wuhan City. It can create favorable conditions for the fundamental management of the Dongting Lake area. In terms of power generation, the Three Gorges Hydropower Station generates an average of 84.68 billion kW·h annually, mainly for East China and Central and South China power supply. In terms of shipping, the 660 km long navigation channel from Yichang to Chongqing of the Yangtze River can be significantly improved, and the 10,000-ton fleet can reach Chongqing port directly, and the one-way annual passage capacity of the channel can be increased from 10 million tons to 50 million tons; the minimum flow in the dry season of the lower reaches of Yichang can be increased from 3,000 m³/s to more than 5,000 m³/s, so that the shipping conditions in the middle and lower reaches of the Yangtze River in the dry season can also be greatly improved. In addition, the project also has great comprehensive benefits such as reservoir breeding, tourism, water supply, and irrigation, as well as the water supply in the long term of South-North Water Transfer (Central Line).

SANMENXIA WATER CONSERVANCY HUB PROJECT

The First Large Water Conservancy Hub on the Main Stream of the Yellow River

Sanmenxia Water Conservancy Hub is located at the junction of Sanmenxia City, Henan Province, and Pinglu County, Shanxi Province, on the main stream of the Yellow River. It is the first large water conservancy hub on the main stream of the Yellow River, with its main task of flood control, anti-logging, irrigation, and power generation.

Panorama of Sanmenxia Water Conservancy Hub after reconstruction

The hub is located in the canyon river section below Tongguan. It is named Sanmenxia (Three Gates Gorge) because there were two islands in the riverbed, which divided the river into three crossing gates. The project started in April 1957, the dam was basically completed in 1960, and the water was stored in the sluice in September of the same year, and the first unit was installed in 1962.

The watershed above the dam site covers an area of 0.6884 million km², with an average multi-year flow of 1,350 m³/s, an average multi-year runoff of 41.9 billion m³, and an average sand content of 37.6 kg/m³. The total reservoir capacity is 15.935 billion m³.

The main buildings of the hub include a concrete gravity dam and powerhouse at dam-toe. The top elevation of the concrete gravity dam is 353 m, the maximum height of the dam is 106 m, and the maximum flow rate of the spill hole of the dam is 6,000 m³/s. The original design of the power station is eight units, with a total capacity of 1,160 MW.

Due to insufficient knowledge about the sediment and reservoir siltation law of the Yellow River, the original design has some defects and results in some problems. After the reservoir was impounded, the reservoir was seriously silted. Despite reducing the water level operation, flood detention, and sand discharge, the siltation continues, and the loss of storage capacity is increasing with each passing day. So, in 1964, the first reconstruction of the project was carried out, and two additional flood discharge tunnels were built on the left bank, and four power generation and water diversion steel pipes were changed to be used for drainage and sand discharge. Although the reservoir siltation has eased, the hub flood discharge scale is still small, failing to solve the general flood siltation problem. After that, two more project reconstructions were carried out in the early 1970s and 1980s, and now the total discharge volume of existing drainage buildings at the hub is 9,701 m³/s.

After reconstruction, the Sanmenxia water conservancy project has played a comprehensive utilization benefit in flood control, ice flood control, irrigation, and power generation. Maintaining the flood control reservoir capacity below the elevation of 335 m, about 6 billion m³, when a large flood of more than 22,000 m³/s occurs in Huayuankou, the hub can partially or fully close the gate control, reducing the burden of downstream flood control; It also relieves the downstream ice jam hazard during the ice-flood period, irrigates 200,000 hm² of farmland, and supply water to the Shengli oilfield and downstream cities; and generates power with a multi-year annual average capacity of 1 billion kW·h.

The practice of the Sanmenxia Water Conservancy Hub Project has deepened people's understanding of the laws of water and sand in the Yellow River, proposed various alteration measures, explored the storage and clearing of mud, water and sand transfer, and other control operation methods. Reservoir siltation has been controlled, and comprehensive utilization benefits are ensured. It also enriched and developed the science of reservoir sediment and accumulated experience for the development and utilization of sediment-rich rivers.

XIAOLANGDI WATER CONSERVANCY HUB PROJECT

Critical Project for the Management and Development of the Yellow River

Xiaolangdi Water Conservancy Hub is located 40 km north of Luoyang City, Henan Province, 130 km from Sanmenxia Dam. The hub is mainly responsible for flood control, silt reduction, taking into account water supply, irrigation, and power generation. It adopts storage of clean water and drainage of muddy operations, eliminating harm and promoting profit so as to make full use of it.

Xiaolangdi Water Conservancy Hub Project is the only control project of the main stream of the Yellow River below Sanmenxia that can obtain a large reservoir capacity. It is the critical project for the management and development of the Yellow River, listed as a national "Eighth Five-Year Plan" key project. The project is strategically important, with a complex project scale, special water and sand conditions, and strict application requirements. So it is regarded by Chinese and foreign water experts at home and abroad as one of the world's most complex, the most challenging water conservancy project.

Xiaolangdi Water Conservancy Hub panorama

The construction of the Xiaolangdi Project started in September 1994 and was completed in 2001. Its completion will effectively control the Yellow River floods, making the flood control standards of Huayuankou on the Yellow River downstream from a 60-year return period flood to a 1,000-year return period flood, basically lifting the threat of flooding in the lower reaches of the Yellow River, slowing down the silting of the lower river. Xiaolangdi reservoir can also use its long-term effective reservoir capacity to regulate non-flood runoff and increase urban and industrial water supply, irrigation, and power generation. It is in the key part of the downstream water and sand control, controlling 100% of the Yellow River sediment discharge.

The dam site controls a watershed area of 0.6942 million km², with an average multi-year flow of 1,342 m³/s and an average multi-year sediment discharge of 1.351 billion t. The normal storage level of the reservoir is 275 m, with a corresponding storage capacity of 12.65 billion m³, of which 7.55 billion m³ is silted sand. The design flood standard of main hydraulic buildings is a 1,000-year return period flood, and the calibration flood standard is a 10,000-year return period flood. The installed capacity of the hydropower station is 1,800 MW, with an average annual power generation capacity of 5.1 billion kW·h for many years.

The hub mainly includes three major parts: a water retaining dam, a flood and sand discharge system and water diversion, and a power generation system. The dam is a clay-inclined core wall rockfill dam with a maximum height of 154 m, which is the largest volume and deepest base cover of the earth's impermeable body local material dam that has been built in China. The total filling volume is 51.85 million m³, the concrete impermeable wall at the base of the dam is 1.2 m thick, the maximum depth is 81 m, and the top is inserted into the inclined core wall for 12 m. The upstream cofferdam is part of the main dam. Under the inclined wall, plastic concrete cutoff wall and rotary spray grouting are combined to prevent seepage. A complete anti-seepage system is composed of the inclined core wall of the main dam, the climbing inner cover, the inclined wall of the upstream cofferdam, and the silting body in front of the dam.

Due to the constraints of topography, geological conditions, and the requirements of siltation prevention at the water inlet, the flood discharge, sand drainage, and water diversion power generation buildings are arranged on the left bank, forming the characteristics of concentrated arrangement of water inlet, cavern group, and outlet dissipation pond. The inlets of 9 flood discharge and sand discharge caves, six water diversion and power generation caves, and one irrigation cave are combined into ten inlet towers lining up, with a maximum height of 113 m. The inlets of each cave are staggered to form a general layout of high-water flood discharge, low-water flood discharge and sand discharge, and middle water diversion and power generation, which can effectively prevent the inlet from siltation. It can effectively prevent inlet blockage, reduce the flow velocity and flow channel abrasion, and improve the reliability of the gate operation. At the outlet, a 2-stage force dissipation pond with a total width of 356 m, a total length of 210 m, and a maximum depth of 28 m were set up to concentrate energy dissipation on the above ten streams, and the drainage channel was connected with the downstream river. Comprehensive management measures of load shedding, drainage, and more than 1,100 prestressed anchor cables' supporting and vertical anti-slip pile reinforcement are adopted to ensure the stability of the high slope. Good results have been achieved.

The water diversion power generation system consists of a power generation inlet tower, water diversion cavern, pressure steel pipe, underground plant, main change room, tailgate chamber, tailwater cavern, tailwater canal, and siltation prevention gate. The overlying rock body of the underground plant is 70–110 m thick, including four layers of mud and a chemical interlayer. A total of 325 prestressed anchor cables of 25 m in length and 1,500 kN in weight are used to support the plant, and prestressed anchored rock wall crane beams are also used in the plant. Six 300 MW hydro generator sets are installed in the plant, and an anti-siltation gate is set at the end of the tailwater channel to prevent muddy water from siltation back to the tailwater cavern during the shutdown. The hydraulic turbine has a large head variation, good hydraulic characteristics, and anti-wear performance. And the setting of a barrel valve can adapt to the sediment-rich conditions and peak regulation operation, and can also carry out maintenance of the water guide mechanism and runner overflow surface without lifting out the rotor and runner.

In the construction process of the Xiaolangdi Water Conservancy Hub, many experiences were accumulated, and many problems were solved, such as the deep cover anti-seepage, inlet anti-siltation, high-speed sandy water dissipation and anti-abrasion, dense cavern group of surrounding rock stability, high slope stability of the entrance and exit of the drainage building, large and complex steel gates, the manufacture and installation of opening and closing machines, etc. A number of innovative technologies were adopted in the design and

Xiaolangdi flood release

construction, such as energy dissipation in the spillway hole with three orifice plates reconstructed from the diversion tunnel, the non-adhesive prestressed concrete lining of the three sand drainage caverns, the GIN grouting, and flat plate joints under the protection of transverse slot hole-filled plastic concrete used in the construction of the seepage control wall, all of which were successful. Xiaolangdi dam filling set many new records in the 20th century in China's earth and rock dam construction: the highest annual strength of 16.361 million m³, the highest monthly strength of 1.58 million m³, and the highest daily strength of 67,000 m³.

DANJIANGKOU WATER CONSERVANCY HUB PROJECT

South-North Water Transfer Central Water Source

Danjiangkou Water Conservancy Hub is located in the jurisdiction of Danjiangkou City, Hubei Province, China, 800 m downstream of the confluence of the Han River and Dan River. It is the first controlling large-scale backbone project for the development of the Han River, and is the water source of the national South-North Water Transfer Central Project, as well as a national-level water source protection zone, an important wetland protection zone in China, and a national ecological civilization demonstration zone.

Danjiangkou Water Conservancy Hub panorama

The Han River has been flooded frequently since ancient times, and by the eve of the founding of the People's Republic of China, it had reached a serious situation of "two collapses in three years and nine disasters in ten years." In 1953, Chairman Mao Zedong inspected the Yangtze River and, after hearing a report on the flood control work of the Yangtze River, included "Danjiangkou" in the great idea of "South-North Water Transfer." In 1954, after hearing a report on the Three Gorges Project, Mao supported the idea of building the Danjiangkou Water Conservancy Hub Project first to train the team and gain experience for the construction of the Three Gorges Project. "Controlling the flood of the Han River," "Training teams for the Three Gorges," and "Transferring water to the north" have thus become the important mission of the Danjiangkou Hydraulic project. In 1958, the Danjiangkou Hub Project was started. The project was developed and constructed in two phases. On September 1, 1958, the opening ceremony of the initial project was held a full month earlier than the scheduled start date. During the extraordinary years of the "Great Leap Forward," "Three Years of Difficulties," and "Cultural Revolution," 100,000 workers carried the project on their shoulders, intercepting the Han River entirely by manpower. The construction process of the Danjiangkou project has had many twists and turns. After the suspension, demonstration, disposal, and resumption of work, the dam was launched in 1967 to store water. At the end of 1973, the hub project was fully completed, playing the role of flood control, power generation, irrigation, shipping, and farming, which was called by Premier Zhou Enlai "all five benefits."

The hub is composed of main buildings such as the left bank earth and rock dam section, the left bank connecting dam section, the plant dam section, the overflow dam section, the deep hole spillway dam section, the ship lifter, the right bank connecting dam section, and the right bank earth and rock dam section. The total length of the water retaining structure is 2,468 m, of which 1,141 m is a concrete dam with wide slit gravity. The left bank earth dam is 1,223 m, with a maximum height of 56 m, a clay core wall and sloping clay wall, gravel material dam shell earth, and stone mixed dam. The left bank connection section between the left bank earth and rock dam and the concrete dam in the riverbed is 220 m long and is a solid gravity dam. The right bank earth dam is 130 m long and is a mixed earth and rock dam with a clay core wall and weathered ballast dam shell. The right bank connection section is 339 m long and is a solid gravity dam. The top elevation of the main dam is 162 m, and the maximum dam height is 97 m.

The basin area above the dam site is 95,217 km², with an average annual runoff of 37.8 billion m³ and an average annual flow of 1,200 m³/s. The design flood standard is a 1,000-year return period flood, and the calibration standard is a 10,000-year return period flood. The design flood flow is 64,900 m³/s, and the corresponding reservoir level is 159.8 m; the calibration flood flow is 82,300 m³/s, and the corresponding reservoir level is 161.3 m. The normal storage level of the reservoir is 157 m.

Six vertical shaft mixed-flow turbine generators with a single capacity of 150 MW are installed in the post-dam type plant, and the water is diverted to generate electricity through six 7.5 m diameter pressure steel pipes buried in the dam. The dam section of the plant is 174 m long and located at the left part of the riverbed. The turbine has a runner diameter of 5.5 m, a rated speed of 100 r/min, a rated output of 154 MW, and a maximum efficiency of 92.8%. The stator core has an inner diameter of 12.8 m. 220 kV and 110 kV out-of-house switching stations are located on the downstream terrace on the left bank.

The navigable structure is arranged on the right bank. The upper section adopts vertical ship lifts with a maximum lifting height of 50 m and a design load capacity of 150 t. The lower section is an inclined ship lift with a track length of 350 m and a slope ratio of 1:7. Together with the upstream and downstream guide walls, the total stroke of the dam crossing facilities is 1,166 m, with a designed annual dam crossing capacity of 830,000 t.

In 1979, the Fifth National People's Congress "Government Work Report" proposed the construction of the South-North Water Transfer Project. In 1987, the Yangtze River Water Resources Commission put forward the "South-North Water Transfer Project Planning Report." In 2000, the central government decided to implement the South-North Water Transfer in three lines: East, middle and west, with an annual transfer of 44.8 billion m³ of water, a total investment of 500 billion yuan, and a construction time

The Danjiangkou Dam after raising

of about 40 years. Danjiangkou Reservoir is the source of water transfer in the middle line. The transfer of water project is divided into two phases of implementation. The first phase of the project began in December 2002, mainly to raise the Danjiangkou Reservoir dam, so that water flows to Henan, Hebei, Beijing, and Tianjin, with an average annual transfer of 9.7 billion m³. On September 26, 2005, the Danjiangkou dam heightening began, with an investment of 2.425 billion yuan, the highest investment in China's water conservancy and hydropower renewal projects. On August 28, 2013, the dam heightening project passed the preliminary technical acceptance before water storage. The height of the dam was raised from the original 162 m to 176.6 m. The normal reservoir storage level was raised from 157 m to 170 m, the reservoir capacity from 17.45 billion m³ to 29.05 billion m³, and the Han River flood control capacity in the middle and lower reaches from a 20-year return period flood to a 100-year return period flood. After the dam was raised, the Danjiangkou Water Conservancy Hub was mainly responsible for flood control and water supply, taking into account power generation, shipping, and ecological protection. 2017 is the Han River basin's abundant water period. Danjiangkou water level for the first time was stored at 167 m, 5 m higher than the original dam top, but the water conservancy hub is in normal operation.

Flood prevention is the top priority of Danjiangkou Water Conservancy Hub: in the 1998 Yangtze River mega-flood, through scientific scheduling, Danjiangkou Reservoir successfully achieved the misalignment of the Han River flood peak with the sixth flood peak of the Yangtze River. By the end of 2018, the Danjiangkou Water Conservancy Hub has intercepted floods with peak flows greater than 10,000 m³/s 92 times, avoiding 12 times of downstream Minwan flood storage and 34 times of Dujiatai flood storage area diversion, mitigating losses to the tune of 62 billion yuan. Water supply is the historical mission of Danjiangkou Water Conservancy Hub: On December 27, 2002, the South-North Water Transfer Central Project officially entered the full-scale restoration and implementation stage. In 2014, the high-quality water from Danjiangkou Reservoir went all the way north, from the Central Plains to the North China Plain, nourishing the farmland along the route and meeting the industrial water and people's daily water needs. By the end of 2018, the Danjiangkou Reservoir has supplied more than 19.5 billion m³ of water, benefiting 53.1 million residents in 19 large and medium-sized cities in the north (11 million in Beijing, 9 million in Tianjin, 15.1 million in Hebei, and 18 million in Henan).

From Mao Zedong's bold vision of "more water in the south and less water in the north, if possible, we may borrow some" in October 1952 to the dream of "a clear water to run Beijing and Tianjin" coming true in December 2014, a history of the planning and construction of the South-North Water Transfer Project has left countless moving and inspiring stories. The history of the South-to-North Water Diversion Project has left countless moving and inspiring stories in China, and nurtured the great spirit of South-North Water Transfer. Danjiangkou Water Conservancy Hub provides a solid guarantee for the safety of the Han River, and has achieved the sharing of resources, becoming a source of happiness that benefits the north and the south, and has successfully completed its mission more than half a century ago, creating a monumental feat in the history of water conservancy.

PANJIAKOU WATER CONSERVANCY HUB PROJECT

China's First Large-Scale Inter-basin Water Transfer Leading Project

In the late 1960s, the city of Tianjin was in severe water shortage due to the drought in North China. In 1973, the State Council decided to build projects to divert the Luan River to Tianjin and Tangshan, with the Luan Diversion Project as the leading one, to alleviate the contradiction between water supply and demand in the Beijing-Tianjin-Tangshan region. Together with the Luan-Tianjin and Luan-Tangshan projects, the Luan River Diversion Project constituted the first major cross-basin water transfer project after the founding of the People's Republic of China. As the source of water for the Luan River Diversion Project, the Panjiakou Water Conservancy Hub Project became the leading project for China's first large-scale inter-basin water supply project to a mega-city.

The Panjiakou Water Conservancy Hub Project is located at the junction of the cities of Tangshan and Chengde in Hebei Province. The dam site is located on the main stream of the Luan River, approximately

Panjiakou Water Conservancy Hub Project panorama

2 km north of Taoyuan Village in Qianxi County. Panjiakou was known as Lulongsai in ancient times and was a key route to the northern and northeastern frontiers in the Ming Dynasty, making it an essential place for soldiers to fight over.

The Panjiakou Water Conservancy Hub Project controls a basin area of 33,700 km^2 and includes the upper reservoir barrage dam, two secondary dams, the overflow dam, the post-dam hydroelectric power station, the lower reservoir barrage dam, the drainage gate, and the riverbed type plant. The upper reservoir, Panjiakou Reservoir, consists of a barrage dam and two secondary dams. With a total reservoir capacity of 2.93 billion m^3, a maximum water surface area of 72 km^2, and a maximum depth of 80 m, the reservoir started construction in October 1975 and passed the handover appraisal and acceptance in June 1985, with a total investment of 0.68 billion yuan. The main dam is a magnificent, wide-slit concrete gravity dam, a class 1 building with 56 sections. The maximum dam height is 107.5 m, the top of the dam is 1,040 m long, the maximum bottom width is 90 m, and the top elevation is 230.50 m. The overflow dam is a top overflow type, arranged in the middle of the riverbed, with 18 table holes, each equipped with a 15 m × 15 m arc gate, and the maximum discharge flow is 531 million m^3/s; in the overflow dam section. In the middle of the overflow dam section, there are four deep drainage holes of 4 m × 6 m in size, also equipped with arc-shaped gates. The two sub dams are earthen dams, with heights of 22.5 m and 5 m, respectively, which do not normally retain water.

The power station plant is a back-of-dam type plant, equipped with three 90,000 kW mixed-flow reversible pumped storage units and one 150,000 kW mixed-flow hydro generator set. The lower reservoir is a pumped storage reservoir formed by a concrete barrage. The dam is 28.8 m high and 1,098 m long at the top, with a normal storage level of 144 m and a minimum level of 139 m, with a total storage capacity of 31.68 million m^3 and an effective storage capacity of 10 million m. There is a riverbed-type plant on the right side, equipped with two bulb-type units with a single capacity of 5,000 kW.

Panjiakou Reservoir main dam

Panjiakou Reservoir's "Great Wall Underwater"

After water storage, the Panjiakou Reservoir's water level exceeded the height of the Great Wall, and the Panjiakou Castle was submerged in water and became the famous "Great Wall Underwater." In September 1983, the reservoir began to supply water to Tianjin, and in December 1984, it began to supply water to Tangshan. Over the next five years, a total of over 4.3 billion m³ of water was supplied downstream, providing Tianjin and Tangshan with a high-quality and stable water source and promoting the rapid development of all sectors of society.

The Panjiakou Reservoir mountain range belongs to the Yanshan Mountain range. Due to the erosion of flowing water, and folding and fracturing effects, the mountains are connected "one after another, one higher than another," the water reflecting the mountain is like scales stacked with brocade. Boating in the green water surrounded by the two-side green mountains, listening to the kingfisher singing on the mountain, people feel like walking in the painting. It is called the "Li River in the north of the country." Panjiakou Reservoir is a large water conservancy hub project to solve the problem of groundwater overdraft in North China and alleviate the shortage of water resources in Tianjin and Tangshan. On average, 1.95 billion m³ of water can be regulated each year and sent to Tianjin and Tangshan to meet the needs of their industrial, agricultural, and urban water needs and to relieve the water supply pressure of Miyun Reservoir, as well as to ensure the safety of the Luan River Bridge of the Beijing-Shanxi Railway downstream. As of August 2020, Panjiakou Reservoir has transferred floods of over 1,000 m³/s 12 times and floods of over 2,000 m³/s nine times. In particular, in July 1994, there occurred the second largest flood ever recorded in the Luan River basin, and the largest since the construction of Panjiakou Reservoir, with a flood peak of 9,870 m³/s. To ensure the safety of downstream flood control, Panjiakou Reservoir exceeded the flood limit by 3.81 m and provided the downstream continuous peak dislocation for 8 h, with flood control and disaster mitigation benefits amounting to 1.7 billion RMB. During the 2012 flood season, under the difficult situation that the flood level of Panjiakou Reservoir was 1.44 m above the flood limit and Daheiting Reservoir was 1.2 m away from the designed flood level, all efforts were made to implement flood peak diversion for the lower reaches of Luan River, which helped to ensure the safety of more than 200 villages, more than 100,000 people and 200,000 mu of arable land in the lower reaches of the Luan River. The completion and operation of the Panjiakou Water Conservancy Hub Project have not only provided water resources for economic construction and social development in Beijing, Tianjin, and Tangshan, but has also played an important role in the economic development and social stability of North China and the whole country.

Manla Water Conservancy Hub Project

Tibet's First Dam

Manla Water Conservancy Hub Project is located on the main stream of the Nianchu River in Longma Township, Jiangzi County, Shigatse City, which is known as "the granary of Tibet." It is 113 km downstream from Shigatse City, 28 km from Jiangzi County Town, and an altitude of 4,200 m above sea level. It is called the "Golden Sun" by the Tibetan people. According to hydrological data, the average annual precipitation of Shigatse station is 421 mm. The inter-annual variation of precipitation is large, with a maximum annual precipitation of 752.1 mm, and a minimum annual precipitation of 210.5 mm. The intra-annual distribution of precipitation is also uneven. The precipitation is mainly concentrated in the flood season, from June to September, accounting for about 90% of the annual precipitation. The runoff of the Nianchu River basin is mainly composed of precipitation, glacial snowmelt, and groundwater recharge. The runoff does not vary greatly from year to year, but the distribution of runoff is uneven within the year, with the flood runoff accounting for 70% of the annual runoff.

Manla Reservoir panorama

On August 26, 1995, the opening ceremony of the Manla Water Conservancy Hub main project was held on the Manla site. On November 20, 1996, the closure of the dam cofferdam was successfully realized. On October 23, 1999, the gate was officially dropped, and water storage started, with the entrance gate of the diversion tunnel successfully lowered. On December 18, 1999, the first unit was officially connected to the grid to generate power. On May 16–18, 2001, the Armed Forces Hydropower Headquarters organized and chaired the preliminary acceptance of completion. The estimated static investment of the Manla Water Conservancy Hub Project is 1,446.3598 million yuan, and the dynamic investment is 1,447.0198 million yuan, which is the largest, most invested with longest construction period project among the 62 aid-Tibet projects determined by the Third Central Conference on Aid-Tibet. It is also the key large-scale water conservancy and energy construction project of the "Brahmaputra and its tributaries Lhasa River and Nianchu River" area's comprehensive development, and therefore known as "the first dam in Tibet."

Manla Water Conservancy Hub power plant

The Manla Water Conservancy Hub Project is a large (II) type project, consisting of a river barrage, a spillway tunnel, a power plant house, and a four-part water diversion system. The dam is a clay wall rockfill dam, with a top elevation of 4,261.3 m, a maximum dam height of 75.3 m, and a top length of 287 m; the reservoir has a total design capacity of 0.157 billion m³, with a normal storage level of 4,256 m, and a dead water level of 4,235 m. The adjusted storage capacity is 83 million m³, and the dead storage capacity is 49 million m³. The average multi-year flow rate at the dam site is 15.1 m³/s, with a 100-year return period, flood peak flow rate of 393 m³/s, and a 2,000-year return period flood, peak flow rate of 644 m³/s. The design flood standard is a 100-year return period, and the design flood level is 4,257.5 m; the calibration flood standard is a 2,000-year return period plus a 20% safety guarantee value.

The calibration flood level is 4,258.4 m, taking into account the upstream glacial lake outburst flood as an extraordinary flood standard for calibration, and the maximum flood level of the dam is 4,260 m.

The power station building consists of the main and secondary plants, the switch station, the pre-plant area, the tailwater canal, etc. The dimensions are 51.5 m × 13.68 m × 25.23 m. The power station is equipped with four 5,000 kW hydrogenerator sets with a designed head of 57.5 m and a total installed capacity of 20,000 kW. The model is HLA562-LJ-135, with a designed annual generating capacity of 61 million kW·h. The water diversion system includes a water inlet, diversion tunnel, pressure regulating well, and pressure pipeline, and the diversion tunnel is 996.13 m long.

The volume of the main project stone excavation is 171 × 104 m³, 104 m³ of cavern excavation, 181 × 104 m³ of dam filling, and 10 × 104 m³ of concrete pouring. The dam site has an elevation of 4,200–4,300 m. The actual construction period is only 7–8 months, and the construction conditions are difficult, with drought and lack of oxygen and severe winter.

Manla Reservoir area

The implementation of the Manla Water Conservancy Hub Project is of great importance to the development of the "Brahmaputra and its tributaries Lhasa River and Nianchu River" area, promoting its economic development, achieving the comprehensive development goals of agriculture and animal husbandry as the mainstay, improving the living standards of the Tibetan people, and gradually narrowing the gap between Tibet and the mainland. The completion of the Manla Water Conservancy Hub Project, known as the "first dam in Tibet," has not only improved the irrigation capacity of the Nianchu River, but has also strongly alleviated the power shortage situation in the Tibetan Central power grid and improved the flood control standards of the three counties and one city downstream of the Nianchu River. During the early days of the People's Republic of China, there were only a few artesian irrigation canals in Tibet, covering only 280,000 mu of irrigated land. The only small power station has been scrapped to stop operation. The urban and rural flood control is basically in a state of non-defense, and drinking water for people and livestock is basically carried by horseback. After the completion of the Manla Water Conservancy Hub Project, the irrigation area will be controlled by 400,000 mu, and the new irrigation area will be increased by 250,000 mu, which will increase the annual grain production by 37 million kg, creating conditions for the Tibet Autonomous Region to achieve grain self-sufficiency. In addition, the project can regulate floods and cut peaks, and raise the flood control standard of the downstream section of the Nianchu River from a 20-year return period to a 30-year return period, which will have great benefits to the role of the granary of Jiangzi, Bailang, and Shigatse. The project stood the test of the 100-year return period flood in August 2000. In 2003, the project won the "Luban Award," the highest honor in the national construction industry, which is the only one in Tibet to receive this award.

BAISE HYDROPOWER WATER CONSERVANCY HUB

The World's First RCC Dam Using Diabase Aggregate

In the Pearl River Basin Flood Control Plan, the Baise Hydropower Water Conservancy Hub is the backbone of the joint effort between the Yu River Embankment and the flood control system. As one of the landmark projects in the water conservancy of Great Great Western Development, it also holds the key to regulating Yu River. The project is located in the upper reaches of the Yu River in Guangxi, with the dam site 22 km away from Baise, a city in the lower reaches. This waterway stretches from Boai in Yunnan to Nanning in Guangxi, covering 435 km, and ranks among the high-grade waterway in the state. By connecting Yunnan, Guangxi, Guangdong, Hong Kong, and Macao, it plays an important role in serving the implementation of the Belt and Road Initiative, advancing the Great Western Development to form new patterns, and battling for poverty alleviation. The Baise Hydropower Water Conservancy Hub is the world's first RCC (Roller-Compacted-Concrete) dam using diabase aggregate.

The State Council approved the project in March 1998, but it wasn't launched until October 1998. The main body part of it started in October 2001, and the river was cut off in October 2002. In July 2006, the first unit of the Baise Hydropower Water Conservancy Hub generated electricity. In November

Baise Hydropower Water Conservancy Hub panorama

2006, the last unit was eligible for commercial operation after the acceptance testing. In total, the project took six years. According to the price level in the first half of 2001, the total investment in the Baise Hydropower Water Conservancy Hub reached 4.713 billion yuan (excluding the transmission costs). The capital funding allocated for the project was 2 billion yuan, shared by the central government and Guangxi provincial government. Among them, the central government shouldered 1.25 billion yuan, while Guangxi provincial government invested 0.75 billion yuan, and the rest 2.713 billion yuan came from loans from the National Development Bank.

Baise Hydropower Water Conservancy Hub is a large-scale project mainly used for flood control. It can also be used for power generation, irrigation, shipping, water supply, and other purposes. The main retaining structures include a major dam and two secondary dams. The main dam is a full-section RCC dam, with diabase as the concrete aggregate. The major dam was 130 m in height, and its crest was 720 m in length and 10 m in width, elevating 234 m above the ground. Its two secondary dams are the Yintun earth and rock dam and the Xiangtun homogeneous earth dam, 39 m and 26 m in height, respectively. The reservoir has a total capacity of 5.66 billion m³ and is used as an incomplete annual balancing reservoir. A space of 1.64 billion m³ is for flood control, while the rest 2.62 billion m³ is for effective storage.

Baise Roller Compacted Concrete Dam (major dam)

Built underground, Baise Hydropower Water Conservancy Hub has an installed capacity of 540,000 kW in total, consisting of four hydro generators, each with a 135 MW capacity. Annually, the average hours of using the power station are 3,130 h, generating an average of 1.69 billion kW·h annually. The adjustable peak power accounts for more than 64%, alleviating the tension between peak and valley time, demand, and supply. With the reservoir in place, the electric power of 9 stages down the stream can be increased by 0.367 billion kW·h in the dry spell. Among them, the completed cascade power stations in Xizu, Guigang, and Guiping, and the cascade power station in Naji, which was under construction simultaneously, will increase the electric power in the dry spell by 0.187 billion kW·h.

The navigation structures are designed for two-stage (2 × 300 t) vertical lifts. The underwater parts of the navigation lock and the main parts are constructed at the same time. The rest of the works are constructed in the second phase. But the facilities for navigation are not completed simultaneously, thus leaving the upper reaches of the You River impassable for nearly 20 years, cutting off the waterway connecting the Guangdong-Hong Kong-Macao Greater Bay Area, which is located in the upper reaches of the Pearl River in Yunnan and Guangxi. In March 2020, the Baise Navigation Facilities Project entity was successfully established, marking a milestone of the preliminary work. Yunnan people's dream of walking from the Pearl River to the sea is about to come true. To accelerate the navigation facilities building, open up the southwest waterway leading to Hainan, improve the overall transportation system in the southwest, and promote the economic and social development of the hinterland, these are the important measures that are to be taken to thoroughly implement important instructions from General Secretary Xi Jinping, battling for poverty alleviation and winning the decisive victory in building a moderately prosperous society in all respects.

Yintun Earth and Rock Dam (secondary dam)

After the completion of the Baise Hydropower Water Conservancy Hub, the outflow of the reservoir will be 9,021 m³/s for 100-year return period floods and 3,000 m³/s for 50-year return period floods. The flood control embankment located in Nanning City is now reinforced according to the standard of a 20-year return period flood. It's expected to resist a 50-year return period flood. Combined with the downstream Laokou Hydropower Water Conservancy Hub, it can resist 100-year return period floods. 1.873 million people and 72,800 hm² of arable land can be protected. The revenues brought by the reservoir can reach 0.775 billion yuan every year.

Baise Hydropower Water Conservancy Hub is known as one of the best projects with the best comprehensive benefits. It has won many prizes, including the "National Excellent Engineering Survey and Design Bronze Award," "National Excellent Water Conservancy and Hydropower Engineering Survey or Design Gold Award," and "China Quality (Dayu) Award for Water Conservancy." Since its commissioning, the project has impounded 12 large floods. The outflow of the floods can exceed 3,000 m³/s, and the maximum peak shaving flow at 4,236 m³/s, remarkably reducing the flood control pressure in Nanning, Baise, and downstream coastal areas. Baise Hydropower Water Conservancy Hub is the main force in the Guangxi power grid. As of April 2020, it has generated 19.009 billion kW·h. In addition to flood control and power generation, the project also benefits farmland irrigation remarkably. It affords sufficient water, irrigating an area of 39,000 hm² from the original 32,500 hm². The irrigation area in hilly land for fruit will be increased by 7,300 hm², and the gravity irrigation area will be expanded by 226.67 hm². In the years 2009 and 2010, the southwest region was hit by a severe drought. With the adjustable reservoir, water was transported downstream incessantly, and the drought in the Yujiang River Basin was eased. Losses were minimalized, and the downstream environment was preserved. In the aspects of social and economic development, Baise Hydropower Water Conservancy Hub has undoubtedly played a remarkable role in promoting the economic development of the Yunnan-Guizhou region and modernizing the old revolutionary areas of the You River.

XINJIANG ULUWATI HYDROPOWER WATER CONSERVANCY HUB

The Small Three Gorges in the Kunlun Mountains

Xinjiang is located in the northwest of China, and Hotan is located in the southern corner of Xinjiang. The Krakash River is one of the important water systems in Hotan, with a length of 808 km and a basin area of 26,600 km². The Uluwati Hydropower Water Conservancy Hub sits at the mouth of the middle reaches of this long and winding river. The Uluwati Hydropower Water Conservancy Hub is a poverty alleviation project. It aims to solve the problems of water and electricity shortage for all ethnic groups in the poverty-stricken Hotan area, known as the "Happiness Project" by the 1.7 million people in Hotan. As the backbone of the Krakash River Basin, it's also the capital project of the national "Ninth Five-Year Plan." This large-scale project promises many benefits: irrigation, flood control, power generation, and environmental protection, known as the "Small Three Gorges in the Kunlun Mountains." Uluwati Hydropower Water Conservancy Hub, a highland reservoir, is also a 4A-level scenic spot deeply located in the desert.

Uluwati Concrete Face Rockfill Dam

The Uluwati Hydropower Water Conservancy Hub is a large (II) type of water conservancy hub. It consists of the major dam, secondary dams, open spillways, flood discharge tunnels on the right bank, flushing tunnels on the left bank, power generation and water diversion cave, major plant behind the dam, auxiliary plant, indoor switch building, and so on. The major dam is 133 m in height, and its crest is 365 m in length. The secondary dam is 67 m in height, and its crest is 96 m in length. The backslope of the main dam has a 10 m-wide zigzag-shaped road up to the dam. Its crest is 89 m in width, elevating 1,967 m above the ground.

The catchment area above the dam site is 19,983 km², accounting for 90% of the basin area, with an average annual flow of 69.5 m³/s. The channel above the Kalakash River dam site stretches for 509 km, controlling 97% of the total river runoff. It is an incomplete annual balancing reservoir, and its total storage capacity is 347 million m³. The normal water level of the reservoir is 1,962 m, and the checked flood level is 1,964.74 m. The capacity below the normal storage level is 323 million m³; the regulation capacity is 225 million m³; the flood control capacity is 28 million m³; the dead storage capacity is 97 million m³. The installed capacity of the power station is 60 MW, with a guaranteed output of 16.5 MW, generating power of 197 million kW·h annually. It also increases the annual effective power generation capacity of the two downstream runoff hydropower stations, namely Karagar and Paizhiwati, by 45 million kW·h.

The flushing tunnel on the left bank consists of an open inlet channel, a gate well for accidents, a pressure cavern, a working gate chamber, a pressureless cavern, an open outlet channel, and an energy dissipation section, with a total length of 811.03 m and a maximum discharge of 123 m³/s. The elevations of the inlet and outlet floor are 1,894 m and 1,879.56 m, respectively. The length of the power generation headrace tunnel is 457.2 m, and the maximum discharge is 90 m³/s, including the open inlet channel, trash rack, emergency gate shaft, low-pressure upper adit section, pressure shaft, penstock, bifurcated pipe, etc.

Open spillway on the right bank

The total length of the flood tunnel is 876.5 m. The inlet and outlet floor elevations are 1,885 m and 1,853.82 m above the ground, respectively. It mainly consists of the diversion channel, accident gate shaft, pressure cave section, working gate shaft, dragon head section and inflow cave combined section, export open channel section, etc. It stretches for 573.85 m, and the maximum discharge is 1,850 m³/s.

The Uluwati Hydropower Water Conservancy Hub began its preparation on August 8, 1993, and it wasn't launched until October 3, 1995. Water flow was cut off in September 1997, and water was impounded in August 1998. 1#, 2# and 3#, 4# Units were put into operation in September 2000. The project began to take full effect in April 2001. It passed the acceptance testing in January 2003. The initial budget was estimated to be RMB 865 million at the 1992 price level, and the revised estimate approved by the State Planning Commission in 1997 was RMB 1.43378 billion. It was jointly invested in and constructed by the State, Xinjiang Uygur Autonomous Region, and Xinjiang Production and Construction Corps. The project excavated 3.63 million m³ of rocks, excavated 157,000 m³ of caverns, consumed 7.12 million m³ of soil and rocks and 340,000 m³ of concrete, and used 2,718 t metal and 8,667 t steel. Main materials: steel 12,246 t, timber 1,782 t, cement 129,611 t, oil 15,765 t, explosives 737 t.

View of Uluwati Reservoir

The Uluwati Hydropower Water Conservancy Hub is a combination of technology, wisdom, and hard work, complemented by nature's extraordinary work. It is the only water control project in Xinjiang that is listed on the national "Centenary Outstanding Civil Engineering," winning the "Dayu Award," "Luban Award," and "Zhan Tianyou Award." This project aims to alleviate the "four hazards and one shortage" in the Hotan area, that is, spring drought, summer flood, salinity, sand, and energy shortage. It benefits society, the environment, and the economy a lot and plays an important role in economic development, ecological well-being, and poverty alleviation for all ethnic groups. In addition to power generation, other benefits are presented as follows:

(1) Irrigation: it has expanded the irrigation areas by 100,000 hm², and improved the irrigation area of 75,000 hm². It can ensure agricultural irrigation, especially in spring droughts;

(2) Flood control: the peak flow of the Krakash River is cut to 500 m³/s for regular floods, and to 890 m³/s for a 50-year return period flood, which is the safe discharge level of the downstream. This has greatly improved the flood control capacity of the river and eliminated the floods downstream;

(3) Ecology: Along with the Hotan River, it can provide an annual water supply of 900 million m³ to the Tarim River and maintain the ecological environment of the green corridor. The Uluwati Hydropower Water Conservancy Hub holds the key to improving the living standards of the Hotan people. We have reason to believe that "Uluwati," a pearl-like project, will be as famous and resplendent as the unparalleled Hotan jade.

ZIPINGPU HYDROPOWER WATER CONSERVANCY HUB

A Project That Stood the Test of the 5.12 Wenchuan Earthquake

The Zipingpu Hydropower Water Conservancy Hub is located in the upper stream of the Min River in Dujiangyan City, Sichuan Province, 9 km from Dujiangyan City and 60 km from Chengdu. As one of the top ten projects of the Great Western Development, the Zipingpu Hydropower Water Conservancy Hub is included in the No. 1 Project of Sichuan Province. The dam was tested by the Wenchuan 8-magnitude earthquake on May 12, 2008. In 2009, the International Commission on Dams and the China Association of Dams awarded it the "International Milestone Award for Rockfill Dam Projects."

The Zipingpu project was listed as a key cascade in the upper reaches of the Minjiang River in 1954. The plan of the project was completed in June 1956, identifying it as the project of the first phase. It is also included in the national "Second Five-Year Plan" key projects, which were among the first key projects supported by the Soviet Union. The project was launched in 1958. However, a gas explosion incurred several casualties when workers excavated diversion caverns. In 1960, the project was shelved due to the technical and economic constraints at that time. At the turn of the century, driven by the national strategy

Zipingpu Hydropower Water Conservancy Hub panorama

of Great Western Development, the Zipingpu project was identified by the state in 2000 as one of the first ten landmark projects. It officially started in March 2001 and has benefited all fronts since then.

The Zipingpu Hydropower Water Conservancy Hub is a large (I) Hydropower Water Conservancy Hub mainly for irrigation and water supply, while taking into account the all-round benefits of power generation, flood control, environment protection, and tourism. The main buildings include a reinforced concrete face rockfill dam, spillway, water diversion and power generation system, sand flushing and emptying tunnel, 1# flood and sand drainage cavern, and 2# flood and sand drainage cavern.

Above the dam site, the controlled basin covered 22,662 km², accounting for 98% of the upstream area of the Min River. Its flow averages 469 m³/s on a yearly basis. The annual runoff is 14.8 billion m³/s, accounting for 97% of the upstream of the Min River, and controlling 98% of the upstream sediment. The project can effectively regulate the upstream water flow, flooding, and sediment. For the checked flood level, it is 883.1 m, and the corresponding flood standard is the maximum possible flood, with a flow rate of 12,700 m³/s. The design flood level is 871.2 m, and the corresponding standard of a 1,000-year return period flood, with a flow rate of 8,300 m³/s. The limited flood level is 850 m. The dead water level is 817 m. The total reservoir capacity is 1.112 billion m³. The reservoir capacity under normal water level is 998 million m³. The flood control reservoir capacity is 166.4 million m³. The balancing reservoir capacity is 774 million m³. The controlled irrigation area is 14 million mu. It's an incomplete annual balancing reservoir.

Zipingpu Hydropower Plant

The dam, a Class 1 building, is a concrete face rockfill dam. The crest of the dam is 634.77 m in length, with an elevation of 884 m. Its wave wall is 1.4 m in height, with an elevation of 885.4 m. The maximum height of the dam is 156 m, with its crest 12 m in width. The foundation of the toe plate has an elevation of 728 m. The maximum depth of the curtain is 110 m.

The spillway is located on the right bank of the ridge. Its chamber is controlled by a positive weir. There's a single hole with an exposed arc gate. The width of the hole is 12 m, and the elevation of the crest is 860 m. The horizontal length of the spillway is 520.5 m, and the maximum flow speed is 42 m/s during flood discharge.

The inlet floor of the sand release tunnel has an elevation of 770 m. The tunnel is located upstream of the inlet section of the diversion tunnel. Its outlet is located downstream of the pick-up section of the spillway, with a diameter of 44 m. The horizontal length of the sand release tunnel is 767.76 m, with a maximum flow speed of 38 m/s during flood discharge. The sand drainage cave is adapted from 1[#] diversion cave and 2[#] diversion cave, respectively. Its shape is like a dragon raising its head. The inlet floor has an elevation of 800 m. The cave section is horseshoe shaped. The diameter of the cave is 10.7 m, and the maximum flow speed is 46 m/s during flood discharge.

The water diversion and power generation system is arranged in the strip ridge on the right bank, including the water inlet, four water diversion tunnels, and the ground plant. The inlet bottom elevation is 800 m, the diameter of the diversion tunnel is 8 m, and the distance between the tunnel axes is 22 m. The main plant is 125 m × 25 m × 54 m (length × width × height), with four hydro generator sets of 190 MW capacity. The power plant is a class 2 building, designed for a one-in-a-century flood and calibrated for 500-year return period floods. The installed capacity of the power station is 760 MW, with a guaranteed output of 168,000 kW, annual power generation of 3.4176 billion kW·h, and annual utilization hours of 4,496 h on average.

In the 5·12 Wenchuan Earthquake of 2008, the Zipingpu Hydropower Water Conservancy Hub was only 17 km away from the epicenter. Its vertical distance is about 5.5 km from the surface rupture in the central fault zone. The basic seismic intensity of the Zipingpu dam site was raised from VII to VIII degrees. At the moment of the earthquake, the dam's power generation system was damaged. All units of

Zipingpu Reservoir

the power station were shut down. The river water could not pass through the units, and the Min River was cut off. The dam's roof and slope suffered varying degrees of damage. Other parts of the hub were damaged to varying degrees. Although the dam was partially damaged, its main body was stable and safe, according to real-time monitoring and assessment by experts from the Ministry of Water Resources. The road collapsed severely after the earthquake. The wide water surface of Zipingpu Reservoir formed a life-saving channel in the water. Within a few days, the Zipingpu Hydropower Station became the first to resume power generation, ensuring water and electricity supply in the downstream area during the tough time. Its existence reassured more than 10 million people downstream. This attested to the scientific siting, reasonable design, and excellent quality of the dam. After the disaster, the station was restored and reconstructed, functioning very well.

JIANGYA HYDROPOWER WATER CONSERVANCY HUB

The Largest Underground Plant In Hunan Province

Jiangya Hydropower Water Conservancy Hub is located in Zhangjiajie City, Hunan Province, China. It's in the middle reaches of the main stream of the Lou River, which is the primary tributary of the Lishui River. Five kilometers downstream from Jiangya Town, 57 km from Cili County, it holds the key to flood control in Lishui. It also enters the list of China's "Ninth Five-Year Plan" key projects and the Ministry of Water Resources and Hunan Province Key Projects. The project is mainly for flood control, but also serves purposes including power generation, irrigation, shipping, water supply, and tourism.

The project was approved in June 1993. But preparations for its construction only began in 1994. The river was cut off at the end of December of the same year. The main project started in July 1995, and all three units were put into operation in December 1999.

Jiangya Hydropower Water Conservancy Hub panorama

The controlled area above the dam site covers 3,711 km², accounting for 73% of the total watershed area. It has a flow of 132 m³/s each year, a measured maximum flow of 6,630 m³/s, and a historical maximum flood flow of 10,000 m³/s. The total reservoir capacity is 1.74 billion m³, with a flood control capacity of 0.74 billion m³. It has irrigated 85,000 mu of farmland, improved the channel stretching 124 km, and provided more than 50,000 people with domestic water.

Jiangya Hydropower Water Conservancy Hub consists of a dam, navigation buildings, a power generation system, and an irrigation water intake system. The dam is a full-section compacted concrete dam. It's 131 m in height. Its crest has an elevation of 245 m, while its foundation has an elevation of 131 m. Among the crushed concrete gravity dams completed or under construction, it is the highest. Its top is 327 m in length. In total, the dame consists of 1.3 million m³ of concrete. The middle of the dam is equipped with a joint flood relief hole in the middle and on the surface.

Jiangya Reservoir has the largest underground plant in Hunan Province at present. It is located in the mountain at the head of the dam on the right bank. The main plant is in the shape of a city gate with a maximum net height of 47 m. The power generation system consists of three diversion tunnels, an underground plant, and a tailwater cave. Three 100,000 kW mixed-flow units are installed in the plant, with a total installed capacity of 300,000 kW and an annual power generation capacity of 756 million kW·h.

Lishui is one of the places that suffered the most from floods in the Yangtze River basin. There are no flood control projects in the whole basin. Historically, there have been many serious floods, which have tortured the people on both sides of the river. In October 1992, the Ministry of Water Resources and the Hunan Provincial Government signed an agreement to harness the Lishui River. Each side agreed to invest 50% in the river basin, so that the whole river basin could be developed on a rolling basis.

The Jiangya Hydropower Water Conservancy Hub is at the core of flood control in Lishui. Once completed, it's expected to resist 20-year return period floods. With over-storage capacity, floods downstream could be reduced to a 30-year return period flood. This has alleviated the pressure of flood control downstream of Lishui and Dongting Lake. It also reduced losses in the middle and lower reaches of the Yangtze River, generated power to western Hunan, and stabilized the power supply grid. At the same time, the Jiangya Hydropower Water Conservancy Hub and the Wulingyuan tourist area are a scenic combination in Zhangjiajie National Forest Park, ranking among the national 4A-level scenic spots. The dam towers over the smoky waters. The high valley and flat lake came together, presenting a visual feast. The water and mountain decorate each other, trees and hills embellish each other, and natural beauty and social customs deepen each other. Many tourists came to Zhangjiajie, promoting local economic development.

LINHUAIGANG WATER CONSERVANCY HUB PROJECT

The Largest Water Conservancy Hub in the Huai River Basin

The Linhuaigang Water Conservancy Hub Project (also known as Linhuaigang Flood Control Project) is located at Linhuaigang, 28 km above Zhengyang Pass, which is known as a "72-water confluence of Zhengyang," in the middle reaches of the main stream of the Huai River, spanning two provinces of Henan and Anhui, involving Gushi County in Henan Province and Huoqiu County, Yingshang County and Fuyang County in Anhui Province. The project is one of the 19 backbone projects of China's Huai River harness, and is also a key project of the national "Tenth Five-Year Plan." It has comprehensive benefits such as flood control, flood removal, optimal allocation of water resources, irrigation, shipping, and tourism, and it is currently the largest water conservancy hub in the Huai River basin.

The project officially started at the end of 2001 and was successfully intercepted one year ahead of schedule in November 2003. The main project was completed and accepted in June 2006. It controls a

Linhuaigang Water Conservancy Hub panorama

watershed area of 42,200 km² and is a large (I) type first-class project, designed according to the standard of 100-year return period flood, with a detention of 8.56 billion m³, and calibrated according to the standard of 1,000-year return period flood, with a detention of 12.13 billion m³. Under the design flood situation, the stagnant storage capacity can reach 8.56 billion m³.

The pivot buildings mainly include the main dam, the north, and south secondary dams, the river diversion, the ship lock, and the floodgate. The entire length of Linhuaigang Dam is 77.51 km and is known as the "world's first dam." The main dam is 8.54 km long, with a top width of 10 m and a top elevation of 31.7 m. The southern secondary dam is of homogeneous earth, 8.41 km long, with a top width of 6.0–8.5 m and a top elevation of 32.25 m. The maximum dam height is 11 m. The northern secondary dam is 60.56 km long, with a top width of 6 m, with a top elevation of 32.21–32.95 m. The dam top is equipped with a 4.5 m wide mud and gravel flood control road.

The inlet floodgate includes 49 holes of shallow hole gate, 12 holes of deep hole gate, and 14 holes of Jiangtang Lake gate, which is among the key Linhuaigang flood control projects. Forty-nine holes of shallow hole gate have a designed flood level of 28.21 m, and the designed flood level under the gate is 26.75 m. And together with other water release structures, it controls the discharge flood of 7,362 m³/s. Twelve holes of deep hole gates are located on the main dam, with a net width of 8 m per hole. Between the original deep-hole gate and the 49-hole shallow-hole gate, the gate has a designed flood level of 18.41 m above the dam and 26.75 m below the dam. 14-hole Jiangtang Lake gate, located on the main dam between the old Huai River main channel and the 49-hole shallow-hole gate. Each hole has a net width of 12 m, and the entire width of the gate is 196.82 m, with a design flood flow of 2,400 m³/s.

Linhuaigang Water Conservancy Hub inlet floodgate

The ship locks include the Linhuaigang ship lock and the Chengxi Lake ship lock. The Linhuaigang lock communicates with the upstream and downstream shipping of the Huai River, with a navigation grade of IV, 500 t, and a lock chamber of 130 m × 12 m. The Chengxi Lake lock has a navigation grade of VI, 100 t, and a lock chamber of 108.65 m × 8 m.

The completion of the Linhuaigang project is a milestone in the history of the Huai River Harness and has changed the flooding situation Huai River basin has suffered for years. As the largest water conservancy hub project on the Huai River, it is similar to the Yangtze River Three Gorges Project and the Yellow River Xiaolangdi Project. So they are collectively known as the three major water conservancy projects in China. It raises the flood standard to a 100-year return period of those in the main flood protection zones below the Zhengyang Pass, like Huaibei Dam, Huainan, and Bengbu in the middle reaches of the Huai River. When the Huai River is in flood, it can avoid flood diversion to the north of the Huai River, greatly reducing the inundation area on the riversides, protecting effectively the safety of railways, highways, and other important traffic arteries, the safety of coal mines and power plants, as well as the safety of the collective property of the state, industrial and agricultural production and people's livelihood in the vast areas in the middle and lower reaches of the Huai River, against flooding and reducing the loss of assets. At the same time, the flood prevention

Partial view of Linhuaigang Water Conservancy Hub

Linhuaigang Water Conservancy Hub plant

and mitigation benefits of the project are also remarkable, with the project's economic benefit-to-cost ratio of up to 1.26. It plays an important role in improving the flood control system of the Huai River basin and safeguarding the stability and development of the regional economy and society.

As a national 4A-level scenic spot, Linhuaigang Reservoir Dam has vigorously developed the local tourism industry. In 2009, the Ministry of Water Resources officially approved it as a national water conservancy scenic spot. The Huai River beneath the dam runs for thousands of miles, with the sky and water of the same color, reflecting the scenery of "thousands of miles long dike locking the Huai River, and thousands of kilos heavy gates cutting the dragon-like water." While developing the economic benefits of tourism, the Linhuaigang Water Conservancy Hub Project also provides an important guarantee for the Huai River basin to build a well-off society and promote the coordinated development of population, resources, environment, and social economy, injecting endless vitality into the new round of vigorous rise on both sides of the Huai River.

YELLOW RIVER SHAPOTOU WATER CONSERVANCY HUB PROJECT

China's First Large Cross-Flow Bulb Unit Power Station

Yellow River Shapotou Water Conservancy Hub Project is located in the main stream of the Yellow River, flowing through Zhongwei City, Ningxia Province, 200 km from Yinchuan City, and 20 km from Zhongwei County. It was listed as one of the top ten new projects in the Great Western Development in 2000 and a key water conservancy project of the national "Tenth Five-Year Plan" in the Ningxia Hui Autonomous Region. The hub is mainly in charge of irrigation and power generation. The project started at the end of 2001 and was cut off in March 2004. The sluice was lowered for water storage, and in May 2005, the main part of the project was fully completed, with a total investment of 1.28 billion yuan.

The dam site has a total control irrigation area of 1.34 million mu. Its design irrigation area covers 877,000 mu, whose design flood of a 50-year return period standard, its flood peak reaching 6,550 m³/s, and whose calibration flood a 500-year return period standard, its flood peak flow of 7,480 m³/s. The total reservoir capacity of the hub reaches 26 million m³, the total installed capacity 120,300 kW, and the annual design power generation capacity of 0.606 billion kW·h, which is categorized as a large type II project.

Yellow River Shapotou Water Conservancy Hub panorama

Shapotou Hub is mainly composed of a riverbed power station, drainage gate, dry canal head power station, drainage holes, and sub-dam. Riverbed power station is a low-head runoff power station, using a cross-flow bulb unit, with a normal storage level of 1,240.5 m, the dead water level of 1,236.5 m, regulating reservoir capacity of 9.5 million m³, able to participate in short-term peaking. The hub's total installed capacity reaches 121,500 kW, including four riverbed power stations, a single unit capacity of 29,000 kW. The two banks of the head of the canal power station have installed one unit each, the north stem power station single unit reaching 2,500 kW and the south stem power station single unit at 3,000 kW. The annual power generation capacity of the pivot power station is 0.606 billion kW·h, which has brought about good social and economic benefits.

The riverbed floodgate has six holes with an orifice size of 14 m × 14.5 m. One drainage hole is provided at the head of the north and south trunk power stations to ensure normal water diversion in the trunk canal when the operation of the power station unit is blocked, with an inlet size of 2.5 m × 3.5 m and an outlet size of 2.5 m × 3 m. The south main canal also has a spillway fork pipe, and in the non-diversion period, uses the fork pipe to release water to ensure the smooth flow of the diversion channel. The dam section of the power station is 867.65 m long, with a maximum height of 37.6 m. The left bank earth and stone mixed sub-dam is about 1,000 m long, with a sub-dam height of 15.1 m and a crest elevation of 1,242.6 m, with a total crest length of 1,369.45 m.

The width of the Yellow River at the dam site is about 165–225 m. The right bank is mountainous; the slope is steep; the left bank is a flat-level terrace; the main stream of the river is on the right bank; the main buildings in the riverbed power station and floodgate are arranged on the main river channel. The hub adopts an open channel to guide the flow, and flood standard accords with a 10-year return period flood with a peak flow of 5,860 m³/s. The open channel is located on the left bank, the bottom width is 70 m, the side slope of excavation on both sides is 1:20, the longitudinal slope is 1‰, the elevation of the bottom of the channel is 1,227–1,228 m, the whole section adopts 0.3 m thick concrete lining, the inlet is protected by precast concrete blocks and lead wire cage blocks, and the outlet is equipped with a force dissipation pool.

The Yellow River Shapotou Water Conservancy Hub Project has opened a new era of dammed water diversion in the Weining Plain, improving the ecology and environment of the region, solving the urban and rural water supply and agricultural irrigation problems in the irrigation area, enhancing the guarantee of water resources regulation and water supply capacity, promoting local economic and social development, and generating huge economic and social benefits. Up to 2020, its cumulative clean energy transmission reached 9.024 billion kW·h, and cumulative irrigation and water diversion 8.644 billion mu, becoming one of the comprehensive water conservancy projects of irrigation and power generation with guaranteed water supply and energy. The completion of the Shapotou Water Conservancy Hub has also improved the ecological environment of the surrounding areas. The former yellow sand-filled Shapotou now transformed into a national 4A level tourist scenic area fulfilling the goal of "green into the sand back." Shapotou, once known for global sand control, is embellished and beautiful because of the oases. Now the Yellow River Shapotou Water Conservancy Hub also adds magnificence and glory to Shapotou. A historic chapter, "Rich World of the Yellow River in Ningxia," will be here to write a new glory.

WANJIAZHAI WATER CONSERVANCY HUB PROJECT

The Largest Water Conservancy Project in Shanxi Province

Wanjiazhai Water Conservancy Hub Project is located in the north main stream of the Yellow River from Toketo to the Longkou River gorge section. Its left bank belongs to the Shanxi Province, and the right bank the Inner Mongolia Autonomous Region, Jungeer Banner. Wanjiazhai is the first of the eight gradient projects in the development planning of the middle reaches of the Yellow River, and also the starting point of the Yellow River into Jin Project in Shanxi Province. The project is mainly to supply water combined with power generation, peak regulation, flood control, and ling prevention.

The project started in 1994, successfully intercepted the flow in December 1995, stored water in the sluice gate on October 1, 1998, and connected the first unit to the grid on November 28. The main body was completed in 2000, all units generated electricity, and completion acceptance was finished on June 29, 2002. The dam site controls a watershed area of 395,000 km², with a total reservoir capacity of 0.896 billion m³ and a regulating reservoir capacity of 0.445 billion m³. The power station has an installed capacity of 1.08 million kW and generates 2.75 billion kW·h of electricity annually.

Wanjiazhai Water Conservancy Hub panorama

Wanjiazhai Water Conservancy Hub Project consists of a barrage dam, water release building, post-dam type plant, switching station, yellow water intake, etc. The barrage is a concrete semi-monolithic gravity dam, using low-heat micro-expansion cement to build the dam. The top elevation of the dam is 982 m, the top length is 443 m, the top width is 21 m, the upstream slope ratio is 1:0.15, and the downstream slope ratio is 1:0.7. The drainage building has eight bottom holes, four middle holes, one table hole, and five sand discharge holes. The bottom hole is a pressure short tube type non-pressure dam body drainage hole, arranged on the left side of 5 to 8 dam sections in the river bed, with two holes per dam section, orifice size 4 m × 6 m, the inlet bottom elevation 915 m using the arc door. When the water level of the reservoir is 970 m, the total discharge is 5,271 m³/s. The middle hole is a pressure short pipe type non-pressure dam body spill hole, arranged in the middle of the riverbed, No. 9 and No. 10 dam section, with two holes per dam section, orifice size 4 m × 8 m, inlet bottom elevation 946 m using a flat door, total discharge capacity reaching 2,156 m³/s. The table hole is an open overflow weir, arranged on the left side of No. 4 dam section, orifice net 14 m wide, the top elevation of the dam reaching 970 m, responsible for drainage and discharge of excessive flood. When the reservoir water level is 980 m, the discharge reaches 864 m³/s. The sand discharge hole is a pressure steel pipe built into the dam, located in the dam section of No. 13–17 power station on the right side of the river bed below the power station inlet with an inlet floor elevation of 912 m. The size of the inlet section is 2.4 m × 3 m, with a flat maintenance gate and an accident gate.

Wanjiazhai Dam

The main plant of the power station is a plant behind the dam, 1,96.5 m long, 27 m wide at the top, 43.75 m wide at the bottom, and 56.3 m high, and in order to solve the problem of anti-slip stability, the plant is connected to the dam by the joint force. The water diversion pressure steel pipe of the power station adopts the water diversion method of a single machine and single pipe, and the water diversion of a single machine is 300 m³/s. The plant has six single-unit 180,000 kW turbine generator sets with a rated head of 68 m, a maximum head of 81.5 m, and a minimum head of 51.3 m.

The head of the canal is two diversion tunnels, using a layered water intake structure, with a diameter of 4 m, a centerline spacing of 12 m, and a single cited flow rate of 24 m³/s. The water intake is arranged on the left bank of the barrage on the non-overflow dam section of No. 2 and No. 3. The highest reservoir water level is 980 m, and the lowest reservoir water level is 957 m during the water intake period.

Aerial view of Wanjiazhai Reservoir

The northern main stream of the Yellow River is one of the regions with severe water shortage in China, with an average annual precipitation of less than 500 mm. After the completion of the Wanjiazhai project, the reservoir operation adopts the operation mode of "storing water and discharging dregs," supplying 1.4 billion m³ of water to Inner Mongolia and Shanxi Province and 200,000,000 m³ of water to Jungeer Banner in Inner Mongolia every year. The Yellow River into Jin project takes water from Wanjiazhai hub, and the total annual water diversion is 1.2 billion m³, of which 0.56 billion m³ is supplied to Pingshuo and Datong in Shanxi, and 0.64 billion m³ to Taiyuan.

The introduction of the Yellow River into Jin is the life project of the Shanxi people showing the achievements of the present and the glory of the future. It has greatly alleviated the water shortage problem in Jin and Mongolia, and is of great strategic importance in solving the water shortage in the two provinces and surrounding areas. Wanjiazhai Water Conservancy Hub Project also optimizes the energy structure of the North China power grid, which can promote the economic and social development of the northwest and even the northern region and improve the power grid operating conditions in North

Wanjiazhai Water Conservancy Hub opening gates to regulate flooding and sand discharge

China. At the same time, the project construction process implements soil and water conservation, greatly improving the ecological environment of the hub construction area. From the once "In the high sky the ravine is worrying about no trees on the deserted side of the bird's nest" to the current dam plant and living area with green space of over 80,000 m² and over 100,000 pines and willows, so the slopes on both sides of the Yellow River presents a unique scenery of flowers and blue waves.

JIANGXI DA'AO WATER CONSERVANCY HUB PROJECT

The First National-Level Reservoir Management Unit in Jiangxi Province

Jiangxi Da'ao Water Conservancy Hub Project is located in Xinjiang River, a tributary of the middle reaches of the Shixi River in Shangrao County, Shangrao City, Jiangxi Province, under the Wufu Mountain in the northern Wuyi Mountains where the three provinces, Fujian, Zhejiang, and Jiangxi Province intersect, 46 km from Shangrao City, Jiangxi Province. This project is one of the "Eight Five-Year Plan" key projects and a large (II) type water conservancy hub project. The main task of the project is irrigation, along with power generation, flood control, water supply, tourism, breeding, and other services, all of which produce comprehensive benefits.

The project was started in December 1995, and the dam was put into storage in June 1999, and two units were put into operation on May 30, 2000, to generate electricity. The dam site controls a watershed area of 390 km², with a multi-year average precipitation of 1,983 mm and a multi-year average runoff of 0.567 billion m³, designed to irrigate 343,300 mu of farmland. The total reservoir capacity of Da'ao

Jiangxi Da'ao Water Conservancy Hub panorama

Reservoir is 0.2757 billion m³; the normal storage level is 217 m; the dead water level is 197 m; the flood control limit level is 217 m; the flood control high water level is 217.6 m; the design flood level is 217.85 m; the calibration flood level is 220.52 m. The design irrigation area is 343,300 mu. The hub is designed in accordance with a century return period flood, and is calibrated according to a 2,000-year return period flood.

The hub building mainly consists of a dam, bank spillway, gate chamber, release tunnel, water diversion and power generation tunnel, and plant. The dam is a concrete panel rockfill dam. The dam is a concrete panel rockfill dam, using soft rock material to fill the main dam rockfill body, the maximum dam height is 90.2 m, and the total length of the top of the dam is 423.75 m. The left bank shore spillway, using shaped pick flow nosecone energy dissipation, is 263 m long, with two holes, each 8 m net wide, and a maximum discharge flow of 1,846 m³/s. This project uses the combination of caving and construction diversion tunnel, placing holes on the left bank with a total length of 314 m. The rear part is transformed from the construction diversion tunnel, and the water diversion system is arranged on the right bank, adopting the form of two machines and one hole, and the tunnel diameter is 5 m. The water extraction point of the diversion project is arranged within 1.5 km of the reservoir dam area, using the tower-type stratified water extraction method.

Da'ao Reservoir

The power plant is located downstream of the right bank ridge, with a total installed capacity of 40 MW and an average multi-year power generation capacity of more than 90.82 million kW·h. The total investment in the project is 0.354 billion RMB.

Since its completion, the Da'ao Water Conservancy Hub has been producing a great comprehensive benefit in flood control, irrigation, power generation, breeding, and shipping, raising the downstream flood control standard from a 5-year return period flood to a 20-year return period flood. Its competent unit passed the management assessment and acceptance of the water conservancy project by the Ministry of Water Resources in 2009, becoming the first national-level reservoir management unit in Jiangxi Province. At the same time, Da Au Hydropower Station is also the first power station in Jiangxi Province to apply full microcomputer monitoring system technology. The power station adopts a hierarchical distributed open system structure, industrial Ethernet connection, stable and reliable performance of the selected equipment, complete software functions, easy and intuitive operation, realizing the automation of project operation and management, and improving the operation and management efficiency.

QINGTONGXIA WATER CONSERVANCY HUB PROJECT

China's Only Gate Pier Type Power Station

Qingtongxia Water Conservancy Hub Project (i.e., Qingtongxia Hydropower Station Hub Project) is located at the outlet of Qingtong Canyon in the middle reaches of the Yellow River in Ningxia Hui Autonomous Region, and is the first phase of the large-scale water conservancy project of the Yellow River gradient development since the founding of the People's Republic of China, as well as the key project of the national "First Five-Year Plan" development and construction and the only gate and pier power station in China. It is also known as the "Pearl of the Seashore." The main task of the project is irrigation, combined with power generation, ling prevention, and industrial water use.

Qingtongxia Water Conservancy Hub Project

The project started in August 1958 and successfully cut off the flow in February 1960. The first unit was put into operation in December 1967, and the civil construction was basically completed. Eight units were put into operation and grid-connected in 1976, and one unit was added in 1993. The hub project is designated as a second-class project, and the main buildings are designed according to the second-class buildings. The normal reservoir storage level is 1,156 m, the total capacity is 0.606 billion m³, the reservoir backwater length is 45,000 m, and the area is 113 km². According to the calculation, after ten years of siltation, equilibrium will be reached, becoming a runoff-type hydropower station with daily regulation capacity.

Qingtongxia Water Conservancy Hub consists of a dam, gate pier plant, sub-plant, switching station, floodgate, irrigation channel, and so on. The project is composed of nine units and seven overflow dams, connected to both banks by an earth dam and concrete gravity dam, and the semi-open-air plant is arranged in the overflow dam gate pier. The dam has a total length of 687.3 m, with a maximum dam height of 42.7 m and a crest elevation of 1,160.2 m. It consists of 34 concrete dam sections with different cross-sections, and the earthen rock dam on the east side of the river is connected to the dam's shoulder. The two main irrigation canals in the east and west are arranged as the main body, and the power station uses the gate pier type arrangement as the core.

The water transmission project of Qingtongxia Water Conservancy Hub consists of three major irrigation channels, namely Qinhan Canal, Tanglei Canal, and Donggaogang Canal, irrigating an area of 366,700 hm². On the left bank, the head of the west canal is introduced into the west main canal (i.e,. Tanglei Canal, built in Qing Dynasty), with a diversion height of 1,136 m and a diversion flow of 400 m³/s. On the right bank, the head of the east canal is introduced into the east main canal (i.e., Qin Canal and Han Canal), with a diversion height of 1,136 m and a diversion flow of 100 m³/s. The upstream high canal, with a bottom canal height of 1,151 m, diverts water with a flow of 24 m³/s.

Local map of Qingtongxia Water Conservancy Hub

The total length of the water retaining building front is 591.85 m; from left to right is the sub-plant dam section, 91.5 m long; the overflow dam and gate pier plant dam section is 262.35 m long; the water retaining dam section is 160 m long, the floodgate dam section is 42 m long; the right bank water retaining dam section is 36 m long. The flood discharge facility adopts seven holes applying the spillway surface flow energy dissipation method, the top of the weir elevation being 1,149.4 m^2. Of the seven holes, the orifice size of two holes is 14 m × 8 m, and that of the remaining five holes is 14 m × 7.5 m, the total maximum discharge reaching 3,255 m^3/s. The floodgate has three holes, with a bottom elevation of 1,140 m and an orifice size 10 m × 5.5 m, with the total maximum discharge reaching 2,205 m^3/s. There are eight units with a maximum quoted flow of 1,860 m^3/s.

Qingtongxia Hydropower Station is a semi-open-air plant with one vertical shaft rotating turbine generator set installed in each gate pier, with a total installed capacity of 300,000 kW and an annual power generation capacity of 1.35 billion kW·h, which greatly relieves the power supply problem in the northwest.

Due to the gate pier type power station arrangement, the beam system of the dam top superstructure was increased accordingly, and the post-tensioned prestressed reinforced concrete structure was adopted to reduce its self-weight and speed up the installation construction. The hydraulics downstream of the rolling dam is articulated with a low nosecone face-flow water flow pattern and a reinforced concrete retaining tank on the bedrock surface of the riverbed below the face-flow bottom water roller. The switchyard uses a double-layer arrangement of high-voltage busbars in the upper and lower levels to reduce the footprint of the switchyard.

Aerial view of Qingtongxia Water Conservancy Hub

Since its commissioning, the Qingtongxia Water Conservancy Hub has given full play to the integrated benefits of water retention, irrigation, flood control, and power generation. The construction of the hub ended more than 2,000 years of history of the Ningxia irrigation area without dams to divert water, greatly improving the guaranteed rate of channel water supply and playing a very important role in the irrigation diversion of this area. After the dam is closed, it can control the flooding of the Yellow River in Ningxia Hui Autonomous Region and Inner Mongolia Autonomous Region, and make Ningxia area form a 10 million mu Yellow River plain irrigation network and mountain water irrigation network. At the exit of Qingtongxia Gorge, 80 km from Yinchuan, the irrigation expansion area eventually reached 7 million mu. While the Qingtongxia Reservoir has helped make Ningxia more beautiful and fertile, the scenery in the reservoir area is also fascinating. The waves are majestic, the beaches are lush, the water birds are swimming, and the ancient pagoda group "108 Pagodas" stands on the left bank of the hill, which has been designated as a nature reserve by the national authorities and is a rare tourist attraction in Ningxia, so the main practical and far-reaching strategic significance of the construction of the Qingtongxia Hub Project can be seen.

JIANGDU WATER CONSERVANCY HUB PROJECT

A Large Water Conservancy Hub Connecting
the Yangtze and Huai River Systems

Jiangdu Water Conservancy Hub Project is located in Jiangdu District, Yangzhou City, Jiangsu Province, downstream of the Yangtze River, at the intersection of the Beijing-Hangzhou Canal, the new Tong Yang Canal, and the Huai River into the Yangtze River at the end of the Manda River, connecting the two water systems of Yangtze River and the Huai River. This project combines a set of water transfer, water supply, irrigation, drainage, navigation, power generation and ecological enhancement, and other functions into one large water conservancy hub, serving as the starting point of the Jiangsu River to North project and of the East-Line Project of South-North Water Transfer.

Jiangdu Water Conservancy Hub panorama

The hub project consists of four large electric pumping stations, 12 large and medium sluice gates, three ship locks, two culverts, two fish passages, transmission and substation works, and drainage channels, of which the four pumping stations are equipped with a total of 33 sets of large vertical axial flow pump units with an installed capacity of 53 MW and a design pumping capacity of 400 m³/s. It is the largest electric drainage and irrigation project in China and even in the Far East.

Jiangdu Water Conservancy Hub Project was built in 1961 and completed in 1977. The Yangtze River water pumped by the project directly irrigates 200,000 hm² of rice fields along the Beijing-Hangzhou Canal and the main irrigation canal in northern Jiangsu Province, including Jiangdu, Gaoyou, Baoying, Huai'an, and Funing counties (cities). At the same time, water is lifted through Huai'an, Zhaohe, Liu Shan, Xietai, and other terrace pumping stations to deliver water for drought relief in northern Huai'an. When the northern Su lower river area is threatened by flooding, the Jiangdu drainage

The largest four-station plant in Jiangdu drainage and irrigation station

and irrigation station can pump water from Jiangdu, Gaoyou, and other three counties (cities) to reduce the water level in the polder area, to ensure the safety of the polder dike and high and stable agricultural production.

When there is more water coming from Huaihe River, under the condition of satisfying irrigation and drainage and water for canal navigation, the reversible unit of Jiangdu 3 station can be started to generate electricity (the main pump is in generator condition) by using the relevant control gate, and its tailwater can be used for irrigation or discharged into Yangtze River in Lixia River area. When there is a shortage of water for navigation on the Beijing-Hangzhou Canal and for towns along the canal trunk line, the Jiangdu drainage and irrigation station can be activated to supply water to the canal. The water is replenished to ensure water for shipping and towns, and it can even send river water to Xuzhou City and Lianyungang City.

Since its completion, the project has pumped and diverted river water for an average of 158 days per year, with an average annual pumping capacity of 4 billion m³, making great contributions to the overall progress of the national economy and social undertakings in northern Jiangsu Province. Party and state leaders including Xi Jinping, Jiang Zemin, Zhu Rongji, Wen Jiabao, Wu Guanzheng, Yang Shangkun, and Li Xiannian have visited the project and have given high praise to it, and it

Jiangdu Drainage and Irrigation Station

has also received guests from many countries and regions.

Jiangdu Water Conservancy Hub Project was reasonably planned and designed, with excellent construction quality and standardized scientific management. In 1982 it was awarded the National Quality Project and the National Golden Quality Award. In 2001, it was identified as the "National Water Conservancy Scenic Area" by the Ministry of Water Resources.

ALTASH WATER CONSERVANCY HUB PROJECT

"The Three Gorges Project in Xinjiang"

Altash Water Conservancy Hub Project is located at the junction of Hoshilaf Township, Shache County, Kashgar Region, Xinjiang Uygur Autonomous Region, and Kusilaf Township, Aketao County, Kizilsu Kirgiz Autonomous Prefecture, and is the largest water conservancy hub project currently under construction in Xinjiang, with a total investment of 10.986 billion yuan, a planned reservoir normal storage level of 1,820 m, a maximum dam height of 164.8 m, and a total reservoir capacity of 2.249 billion m³. The total installed capacity of the power station is 755 MW, and the designed annual generating capacity is 2.26 billion kW·h. On October 10, 2011, the official construction was started, and on August 31, 2019, the dam was capped. On November 19, 2019, the Altash Water Conservancy Hub Project reservoir dam began lower sluice water storage.

Aerial view of Altash Water Conservancy Hub Project

The hub project mainly consists of a barrage dam, water discharge building, power generation diversion system, power plant, ecological base-flow discharge cave (ecological power plant), and other buildings. The project can ensure ecological water transfer to the Tarim River, flood control, irrigation, power generation, and other comprehensive use of the function, and meanwhile, after completion, can greatly regulate the distribution of the Yarkant River runoff in the year. As introduced, the project can effectively control the upstream mountain floods, and combine with the construction of the planned embankment to raise the flood control standard of the downstream key protection objects from a less than 20-year return period flood to a 50-year return period flood, while usually the flood control standard of the general protection objects only ranges from a 2.5-year return period flood to a 20-year return period flood. The project can solve the spring drought water shortage in the irrigation area of Yarkant River Basin and the heavy flood control burden of the people in the irrigation area, change the power shortage in the basin and the three southern Xinjiang regions, improve the ecological environment of Tarim River Basin and Yarkant River Basin, and promote the sustainable development of the national economy and society in the basin and the three southern Xinjiang regions.

Due to the complex topography of the area where the main project is located, it is necessary to overcome the "three high and one deep" worldwide problems in design and construction. General water conservancy project, either deep cover, high dam, or high side slope, only encounters one difficult point, but the Altash Water Conservancy Hub Project concentrates on all the difficult points. Therefore, it is called the "Three Gorges Project in Xinjiang" by industry experts. In order to overcome these technical problems, the construction side relied on many scientific research institutions and colleges, established a number of scientific research projects, docked the application of the most advanced construction technology and concepts, and jointly conquered all difficulties. The dam adopts a rare high-slope excavation and support treatment at home and abroad, using the world's leading high-panel rockfill dam technology. As the dam is located on a soft foundation under which nearly 100 meters deep cover layer, and the total height of the dam foundation cover layer and the dam body reaches 258.8 m, concrete piles were driven into the soft foundation more than 90 meters deep. Workers used foundation consolidation grouting and gravel consolidation grouting to reduce foundation settlement and stabilize the dam foundation. Lower sluice water storage means that all the technical solutions have been transformed into engineering practice with the support of scientific research results, marking a new level of scientific research and design for water conservancy construction in Xinjiang.

In addition to power generation, the Altash Water Conservancy Hub Project of the ecological power station also ensures the original river ecological overflow between the dam site to the main plant. In 2019, the Altash Water Conservancy Hub Project completed a fish breeding station, including the Tarim split-belly fish, broad-mouthed split-belly fish, and four other kinds of indigenous fish breeding. In early October 2020, the Altash Water Conservancy Hub Project released 100,000 tail fish seedlings for the first time, mainly releasing four species of indigenous fish in Xinjiang, including Tarim split-belly, spotted heavy-lip, wide-mouth split-belly and thick-lip split-belly, of which Tarim split-belly and spotted heavy-lip are class II aquatic wildlife protected animals in the autonomous region, contributing to the restoration of local ecology.

DAYUANDU NAVIGATION AND POWER HUB PROJECT

The First Navigation Hub Project That Uses Electricity to Support Navigation

Dayuandu navigation and power hub project is located on the main stream of Xiangjiang River in Hengshan County, Hunan Province, 62 km away from Hengyang City, which is the first hub project of the 1,000-ton navigation channel from Hengyang to Chenglingji section of Xiangjiang River to support navigation with electricity, and is also a pilot project of the Ministry of Transportation in the inland river to "combine navigation and electricity, promote navigation with electricity." The project is based on the canalization and improvement of High channel grade, to meet the large fleet of water transport as the main task, while taking into account the benefits of power generation, flood control, irrigation, and tourism.

Dayuandu Avionics Hub panorama

The project consists of a concrete gate dam, riverbed type plant, ship lock, and earth sub-dam. The dam site controls a watershed area of 53,200 km², with an average multi-year runoff of 44.1 billion m³. The normal storage level of the reservoir is 50 m, with a storage capacity of 451,000,000 m³, and the design flood standard of the gate and dam is a 50-year return period flood, and the calibration flood standard is a 500-year return period flood.

The total length of the lock and dam is 533 m, with 23 holes, an exposed top type arc gate, and automatic control hydraulic opening and closing. The maximum height of the secondary dam is 32.5 m. The maximum head of the power station is 11.24 m, which is a medium-sized power station. The plant is located on the right bank, with four sets of 30 MW light bulb cross-flow turbine generator sets, a total installed capacity of 120 MW, an annual average power generation capacity of 0.585 billion kW·h, a maximum operating head of 11.24 m, minimum operating head of 2 m, rated head of 7.2 m. For the first time in rainy areas, the plant's movable roof (movable and manual) is used, with dimensions of 10 m × 15 m and 30 m × 15 m. The power station's dirt cleaning machine integrates dirt cleaning, dirt transmission, and barrage lifting. The ship lock is arranged on the left bank terrace, and the size of the single-stage lock is 180 m × 23 m × 3 m. The total length of the canal is 2,300 m, and the convex bank is cut and straightened. The reservoir storage period navigation break time is 45 days.

The project started on December 20, 1995, and the first unit was connected to the grid at the end of December 1998, and the locks were officially opened to navigation; the fourth unit generated electricity in October 1999 and was completed in May 2000, half a year earlier than the original schedule, creating the two best results of the shortest construction break period and the first unit generation period of the same scale hub project in China.

The completion of the Dayuandu Navigation

Dayuandu Navigation and Power Hub locks

and Power Hub has provided a stable source of funds for further development and utilization of the Xiangjiang River and also set a successful model for diversified financing of inland navigation development. The normal operation of the project marks the great success of the rolling development strategy of "combining navigation and electricity, promoting navigation with electricity" implemented by China in the development of inland waterway navigation.

QUNYING RESERVOIR PROJECT
Asia's Highest Masonry Gravity Arch Dam

Qunying masonry gravity arch dam is located on Dasha River, a tributary of Weihe River, 500 m north of Dahepo Village, Xiuwu County, northwest of Jiaozuo City, Henan Province, with Jiaozhi Railway and Xinji Highway 28 km downstream of the dam. The project is mainly for irrigation and water supply, along with flood control, power generation, and fish farming.

Qunying Water Conservancy Project panorama

The project consists of three parts: a dam, a water transfer cave, and a spillway. The project started in November 1968 and was basically completed in July 1971. After the "75-8" flood, according to the 1,000-year standard calibration, the dam was raised by 5.5 m in 1976. From April 1999 to May 2000, the project was repaired, reinforced, and completed the grouting of the left and right shoulders of the dam, flood control road renovation, and management facilities support.

The dam site controls a watershed area of 165 km², the average multi-year rainfall is 650 mm, the design flood level of the reservoir is 481.75 m, the check flood level is 485.2 m, the total reservoir capacity is 16.6 million m³, the profit-raising capacity is 12.7 million m³, the design irrigation area is 4,000 hm², and the design daily water supply is 48,000 m³. The original design of the power station is 1 MW, which is later modified to 0.75 MW.

The dam site is a steep mountain range. The river valley is V-shaped, the width of the riverbed is 27 m, the topography is basically symmetrical, and the rocks are relatively complete and gentle. The dam is a fixed-center variable-radius slurry masonry gravity arch dam with an arc length of 154.28 m, a center angle of 80°, an outer radius of 110.5 m, a crest elevation of 490.5 m, a maximum height of 101 m, a crest thickness of 4.5 m, a bottom thickness of 52 m, and a thickness-to-height ratio of 0.52, becoming the world's highest dam of its type. The dam is designed for a 50-year return period flood and calibrated for a 1,000-year return period flood. The top of the dam is equipped with seven holes and 8 m wide overflow weir. The total length of the overflow section is 65 m, using the pick flow energy dissipation, and the maximum discharge volume is 2,480 m³/s. The water transfer hole is divided into two levels, both located on the left bank; the elevation is 412 m and 445 m, the inner diameter is 1 m and 0.8 m, and the outlet is controlled by a gate valve.

Qunying Dam

Since the completion of the reservoir, through the joint efforts of leaders at all levels and all water conservancy project managers, the reservoir irrigation and water supply benefits are remarkable, and the comprehensive benefits of water conservancy have been further developed. In 1992 relying on Qunying Reservoir to set up the Jiaozuo City Fenglin Gorge Scenic Area, tourism also gained significant development.

SHIBIANYU RESERVOIR PROJECT

China's Highest Earth and Rock Dam with Impermeable Reinforced Composite Geomembrane

Shibianyu Reservoir Project is located downstream of the Shibianyu River at the northern foot of the Qinling Mountains in Chang'an District, Xi'an City, Shaanxi Province, 35 km from Xi'an City. The project focuses on irrigation and urban water supply, along with other benefits such as power generation and flood control.

Shibianyu Reservoir panorama

The project started in 1972 for the construction of the diversion cave, the left bank of the main explosion area, the right bank of the secondary explosion area of the guide hole, and chamber construction. On May 10, 1973, the directional blasting exploded, with a blasting volume of 2.36 million m³, an upper dam volume of 1.44 million m³, and an upper dam rate of 60.7%, becoming the largest amount of China's directional blasting dam construction. In 1980 the project was basically completed, and in 1981, the first water storage was started.

The dam site controls a watershed area of 132 km², with an average annual runoff of 97 million m³. The total reservoir capacity is 28.1 million m³, with an effective capacity of 26.5 million m³, a design irrigation area of 11,200 hm², and an annual water supply of 30 million m³ to Xi'an City. The design flood standard is a 100-year return period flood, and the calibration flood standard is a 1,000-year return period flood.

The project consists of a dam, water transfer cavern, flood relief cavern, and two-stage hydropower station. The dam is a directional blast rockfill dam with a maximum height of 85 m and a crest length of 265 m. The total volume of the dam is 2.08 million m³, including 1.44 million m³ of the directional blast rockfill dam. Asphalt concrete impermeable sloping wall adopts a simple section without a drainage layer, with thicknesses of 32 cm, 27 cm, and 22 cm, respectively, according to the head. The water transfer cave is located on the left bank, which is used for diversion, irrigation, power generation, water supply, and flood discharge. It is a circular pressure tunnel with a diameter of 4 m, designed for a maximum discharge of 192 m³/s and an irrigation diversion flow of 10 m³/s. The flood relief cave is located on the right bank and is a non-pressure tunnel in the shape of a city gate cave designed for a maximum discharge of 808 m³/s.

After the reservoir was filled with water, leakage occurred in 1980, 1992, and 1993, and the measured maximum leakage amount behind the dam reached 172 m³/s, for which the reservoir was operated at a lower water level. The main reason for the leakage behind the dam is the poor grading of the artificial filling part of the dam, and the original slope of the right bank was not removed when the asphalt concrete sloping wall was paved, resulting in cracks and slumping of the impermeable sloping wall caused by leakage. From January to June 2000, the project took filling and grouting to the shallow rock pile of the dam and reinforced the surface of the asphalt concrete sloping wall with a composite geomembrane, which achieved good results and is the dam with the highest leakage defects treated with a composite geomembrane in China.

THE SOUTH-NORTH WATER TRANSFER PROJECT

The Water Transfer Project Crossing Most River Systems

The South-North Water Transfer Project is the generic name for the project that transfers water to northern China from the lower, middle and upper reaches of the Yangtze River in three routes (eastern, central, and western), which formed a water network pattern of "four-lateral and three-longitudinal routes" interconnected with the Yangtze River, the Yellow River, the Huai River, and the Hai River. After the final completion of the three routes, the preliminary plan was to transfer a total of about 38 billion to 48 billion m³ of water annually, which is close to the water volume of another Yellow River in the Huang-Huai-Hai Plain and the northwestern region. It basically transformed the serious water shortage in North China.

The term "South-North Water Transfer" first appeared in official central party literature in August 1958. At an enlarged meeting of the Political Bureau of the CPC Central Committee held in Beidaihe, the CPC Central Committee approved and issued the "Instructions of the CPC Central Committee on Water Conservancy," which clearly stated that "In addition to the planning carried out in various regions,

The South-North Water Transfer Project Sihong Station

the nationwide long-term water conservancy planning prioritizes the northward transfer of south water (mainly refers to the Yangtze River system) as the main goal, which is a plan of linking the Yangtze River, Huai River, Han, and Hai River basins into a unified water conservancy system." Prior to this statement, when being briefed on the idea of diverting the Yangtze River to the Yellow River on October 30, 1952, Mao Zedong said, "Southern China has more water while Northern China has less. So, if it's possible, it is acceptable for the north to borrow some water from the south."

Under the accurate leadership and care of the Party and the government, numerous science and technology workers conducted more than 50 years of scientific research on South-North Water Transfer continuously. They carried out a large number of field surveys and measurements and formed the basic scheme of eastern, central, and western routes for the South-North Water Transfer based on the analysis and comparison of more than 50 schemes, obtaining large amounts of valuable results.

The project would be implemented in three phases over the next 50 years, with an estimated total investment reaching RMB 486 billion.

The Eastern Route Project: Utilizing the existing Jiangsu-North Water Transfer Project of Jiangsu Province, the project gradually expands the water transfer scale and extends the water transmission line. The Eastern Route Project pumps Yangtze River water from Yangzhou at the lower reaches of the Yangtze River and draws on the Grand Canal and its parallel river channels to transfer water northward level by level. It also connects Hongze Lake, Luoma Lake, Nansi Lake, and Dongping Lake, all of which play the role of water adjustment and storage. After passing Dongping Lake, water transits in two routes: One route goes all the way to the north, crossing the Yellow River through a tunnel in the vicinity of Weishan. The other route goes east, passes through the Jiaodong area water supply trunk line, and delivers water to Yantai and Weihai through Jinan. The Eastern Route Project can provide a net increase of 14.8 billion m³ in water supply for Jiangsu, Anhui, Shandong, Hebei, and Tianjin.

The Central Route Project: The project diverts water from the inlet sluice of Taocha Canal of Danjiangkou Reservoir, of which the capacity was extended with new dams built. Along the west side of the Tangbai River basin, the route first goes past the Fangcheng pass, which is the drainage divide of the Yangtze River basin and the Huaihe River basin. Through the western edge of the Hunag-Huai-Hai Plain, the route continues by crossing the Yellow River in the west of Zhengzhou at the west Gubaizui and going northward along the west side of the Beijing-Guangzhou Railway. It can flow to Beijing and Tianjin mostly by itself. The Central Route Project can alleviate the water crisis in Beijing, Tianjin, and North China, increase the

Danjiangkou Reservoir Dam, the water source for the central route of the South-North Water Transfer

water supply by about 10 billion m³ for Beijing, Tianjin, and cities along Henan and Hebei for living, industry, and agriculture uses, improve the ecological environment and investment conditions in the water supply area greatly, and promote the economic development of the central region of China.

The Western Route Project: The project builds dams and reservoirs in the upper reaches of the Yangtze River at Tongtian River, Yalong River, and Dadu River. It digs through the water transmission tunnel of Bayan Har Mountain, which is the drainage divide of the Yangtze River and the Yellow River, to transfer water from the Yangtze River into the upper reaches of the Yellow River. The water supply goal of the Western Route Project is to solve the water shortage problem in the upper and middle reaches of the Yellow River and the Guanzhong Plain of the Wei River, which involves six provinces, including Qinghai, Gansu, Ningxia (autonomous regions), Inner Mongolia (autonomous regions), Shaanxi, and Shanxi. Together with the construction of the major hydropower junction project on the main stem of the Yellow River, it can also supply water to the Gansu Hexi Corridor Area adjacent to the Yellow River basin and, if required, the lower reaches of the Yellow River. The three rivers included in the Western Route Project can transfer water of about 20 billion m³, develop irrigation areas of up to 2 million m² for six provinces, including Qinghai, Gansu, Ningxia (autonomous regions), Inner Mongolia (autonomous regions), Shaanxi, and Shanxi, and provide 9 billion m² water for urban living and industry uses. It can also advance the economic development of northwest inland areas and improve the ecological environment of the Loess Plateau in northwest China.

The total water transfer volume for the planned eastern, central, and western routes will be 44.8 billion m³ by 2050. 2002–2010 is the early stage of the South-North Water Transfer Project, with a total water transfer volume of about 20 billion m³. 2011–2030 is the middle stage, with an increase in water transfer volume of 16.8 billion m³ and an accumulated total volume of 36.8 billion m³. 2031–2050 is the final stage, with an increase in total annual water transfer volume of about 8 billion m³.

The Eastern Route Project of the South-North Water Transfer officially started on December 27, 2002. The Crossing Yellow River Project tunnel of the Central Route was officially kicked off on September 27, 2005, and the Henan section officially started on September 29, 2006. At present, the central route and the eastern route (Phase I) projects of the South-North Water Transfer Project have been completed and transferred water to the northern region. The western route is still in the planning stage, and its construction has not started.

(1) The Phase I Project of the Eastern Route is 1,467 km long. Its main project consists of three components: the water supply project, the water storage project, and the power supply project. Along the whole line, there are 13 cascade pumping stations and a total of 22 hubs with 34 pumping stations. In total, it has a 65 m long hydraulic head, 160 installed units, an installed capacity of 366,200 kW, and an installed flow of 4,447.6 m³/s. It is characterized by its large scale, numerous pump types, low head, high flow rate, and high annual utilization hours. The Eastern Route Project created the world's largest pumping station group—Eastern Route Pumping Station Group Project. The project is implemented in three phases, with a total of 21 additional pumping stations built in the first phase, 13 additional pumping stations built in the second phase, and 17 additional pumping stations built in the third phase. After the completion of the project, it will become a modern pumping station group with the highest number of concentrated large pumping stations in Asia and the world. Its hydraulic model of pumps and pump manufacturing level both reached the advanced international level.

(2) The important part of the Central Route Project is to heighten the dam of Danjiangkou Reservoir. Its water supply project is mainly based on open channels, with a combination of pipeline water

transmission pressurized by pumping stations in certain canal sections. It has a total trunk canal length of 1,432 km. In the afternoon of December 12, 2014, the 1,432 km long central route of the South-North Water Transfer Project, which took 11 years to build, was officially opened, bringing Yangtze River water to Beijing. The water quality of its water source Danjiangkou Reservoir is maintained above national water quality Class II year-round with a "double sealed" channel design to guarantee water quality safety along the route. After the opening, the route can deliver 9.5 billion m³ of water to the north every year, which is equivalent to 1/6 of the Yellow River. As of June 3, 2020, the first phase of the middle line of the South-North Water Transfer Project has safely delivered water for 2,000 days, with an accumulated total volume of 30 billion m³ delivered to the north, benefiting 60 million people along the route. The water supply safety factor of the central area of Beijing has thus been raised from 1 to 1.2. The shallow groundwater level in Hebei Province has grown from a 0.48 m increase per year before the opening to a 0.74 m increase per year.

On October 23, 2020, China South-North Water Transfer Company was officially founded. It is a wholly state-owned new central enterprise with a tentative registered capital of 150 billion yuan. The investment in South-North Water Transfer Project has surpassed the Three Gorges Project, making it the world's largest water conservancy project.

DONGJIANG-SHENZHEN WATER SUPPLY PROJECT

The World's Largest Water Transfer and Purification Project by the End of the 20th Century

The Dongjiang-Shenzhen Water Supply Project is located in Dongguan City and Shenzhen City in Guangdong Province. It is an inter-basin large-scale water transfer and purification project that mainly targets Hong Kong and provides drinking water and irrigation water for farmland in Shenzhen and towns in Dongguan along the project route as well.

The total length of the water transmission project is 83 km, of which the water intake is located on the right bank of Dongjiang River in the Pearl River system in Qiaotou Town, Dongguan City. After the water is lifted at the Taiyuan pumping station, it will be diverted by channel to the Sima pumping station and then diverted by channel again to the Shima River and Yantian Water River. Along the river, barrages of five stages have been set up in reverse directions at Qiling, Matan, Tangxia, Zhutan, and Shaling. Four pumping stations have been set up near the barrages at Matan, Tangxia, Zhutang, and Shaling, which

Shenzhen Reservoir panorama

allow the water to flow back along the Shima River and Yantian Water River with a four-stage lift. Water then enters the Shenzhen Reservoir after the 6 km long Yantian tunnel with biological pretreatment. Finally, the water is released from the left sub-dam of Shenzhen Reservoir and sent to Hong Kong via four hydroelectric power stations with an installed capacity of 1.6 MW and pressure steel pipes.

The project was completed and put into operation in March 1965, with an annual design water supply capacity of 68 million m³. With the socio-economic development of Hong Kong and the towns along the route, the water demand has been increasing. At the request of the HKSAR government, the project was expanded three times in the 1970s, 1980s, and 1990s, with the total annual design water supply capacity increased to 1.743 billion m³, of which the capacity for Hong Kong and Shenzhen is 1.1 billion m³ and 493 million m³ respectively.

The Qiling Pumping Station of the Dongjiang-Shenzhen
Water Supply Project

The world's largest U-shaped aqueduct

The biological pretreatment project, also known as the biological nitrification project, is located at the tail area of Shenzhen Reservoir, which is the end of the water transfer project. It is an important component of the Dongjiang-Shenzhen Water Supply Project, which was completed and put into operation in December 1998. The project is responsible for removing a variety of pollutants in raw water, mainly ammonia and nitrogen, to deliver high-quality raw water for Hong Kong and Shenzhen. The design volume of water treatment of the project is 4 million m³/d, making it the world's largest raw water pretreatment project by the end of the 20th century. The main body of the project, a biological contact oxidation pool, is made up of six pools with effective lengths, widths, and depths of 270 m, 25 m, and 3.8 m, respectively. Each pool is set up with twenty 12 m long, 25 m wide, and 3 m high filler squares using elastic tridimensional filler. At the bottom of the pool is a perforated pipe aerator, of which air is supplied by six blowing engines, each with an air volume of 555.6 m³/min. There is a floodgate on the right side of the pretreatment pool.

The Dongjiang-Shenzhen Water Supply Project has solved the long-term freshwater shortage in Hong Kong. The water supply volume of the project accounts for about 80% of the total water consumption in Hong Kong and has become an important factor in ensuring the stability and prosperity of the Hong Kong region. It also plays an extremely important role in the rapid economic development of the Shenzhen Special Economic Zone and areas along the Dongjiang-Shenzhen route.

LUANHE RIVER-TIANJIN WATER DIVERSION PROJECT

An Inter-basin Urban Water Supply Project

The Luanhe River-Tianjin Project is an urban water supply project that diverts water from the Luanhe River in Hebei Province to Tianjin across river basins. Its water source is located at the Panjiakou Reservoir in the middle and lower reaches of the Luanhe River in Qianxi County, Hebei Province, which provides an annual water supply of 1 billion m³ to Tianjin at 75% of its design dependability.

The project started on May 11, 1982, and was completed and put into operation on September 11, 1983, after 16 months of construction. It is a model of high-speed construction of large-scale water transfer projects in China in the 1980s.

Panjiakou Reservoir—the water source for the Luanhe River-Tianjin Water Diversion Project

The water of the project is released from Panjiakou Reservoir and adjusted at Daheiting Reservoir along the Luanhe River. The water diversion junction is at the starting point of the Luanhe River-Tianjin Water Diversion Project. After crossing the drainage divide, water enters the Yuqiao Reservoir in Tianjin along the Li River in Zunhua City, Hebei Province, for adjustment and storage through the diversion tunnel. It then goes south along the Zhou River and the Ji Canal River to enter the special open channel for water transfer, where it is lifted and pressurized into the Hai River. Water is then transmitted into three water treatment plants—Jie Yuan, Ling Zhuang, and Xinkai River, via buried culverts and steel pipes. The total length of the water diversion route is 234 km.

The entire project consists of 215 projects, such as a water diversion junction, water diversion tunnel, river regulation project, Yuqiao Reservoir, Er'wangzhuang Reservoir, pumping station, water transmission open channel, and canal structures.

The water diversion junction contains two sluice gates for transmitting water into the Tianjin and Tangshan areas in Hebei Province, with a diversion flow of 60 m³/s and 80 m³/s, respectively.

The total length of the project, including the water diversion tunnel, the entrance, and the exit, is 12.39 km, of which the tunnel is 9.66 km long. The tunnel is built in round-arch and straight-wall style with a net width of 5.7 m and a net height of 6.25 m. It passes through a 212 m-long rare mega fault along the route. To guarantee project quality, the project draws on the advanced experience of contemporary underground engineering design and construction, adopts the New Austrian Tunneling Method, and incorporates actual new design and construction technology into its construction.

The project regulated river channels 108 km long in total, evacuated open channels 64 km long for water transmission, and built twelve inverted siphons, five culverts, and seven sluices.

With the economic development of Tianjin and the improvement of people's living quality, the main project of the Luanhe River-Tianjin Water Diversion Project has gradually distributed water for supply, with six expanded water supply branches, and trunk lines, prestressed concrete pipes with a total length of 414 km, eight newly-built pumping stations, and an annual water supply increase of 258 million m³. As of 2000, the project has delivered 14.7 billion m³ of water to Tianjin, which has created great social, economic, and environmental benefits.

IRRIGATION PROJECT OF DIVERTING DATONG RIVER TO QINWANGCHUAN

China's Largest Inter-basin Water Diversion for Gravity Irrigation Project

Diverting Datong River to Qinwangchuan is a large-scale self-flow irrigation project supported by inter-basin water transfer in Gansu Province. It transfers water from the Datong River, which originates in Qinghai Province, to the Qinwangchuan area about 60 km north of Lanzhou City. The project is mainly developed for settling 80,000 immigrants in the poor areas of eastern Gansu Province and solving the production and domestic water of 400,000 people in the irrigation area as well as improving the ecological environment of Qinwangchuan area, gradually increasing the vegetation cover and forming an ecological barrier in the north of Lanzhou City, which creates outstanding economic, social and environmental benefits.

The canal aqueduct of "Diverting Datong River to Qinwangchuan" project

The project started construction in 1976, and the main project was completed and launched in 1995. The annual design self-flow water diversion volume of the project is 443 million m³, with a total irrigation area of 58,700 hm² and an annual grain yield increase of about 150 million kg. The project consists of a diversion canal head area, a water transmission canal system with its structures, and coordinated field projects. The general main canal is 86.94 km long, extending from the head of the Tiantang Temple diversion canal to the general diversion sluice gate at the Xianglu Mountain in Yongdeng County, Gansu Province. The design diversion flow of the trunk canal is the same as the intake sluice at the canal head. At the diversion sluice gate of Xianglu Mountain, water is diverted into the East First Trunk Canal, East Second Trunk Canal, and 45 branch canals to flow into the irrigation area. The East First Trunk Canal is 52.66 km long, with a design diversion flow of 14 m^3/s and an irrigation area of 21,100 hm^2. East Second Trunk Canal is 53.62 km long, with a design diversion flow of 18 m^3/s and an irrigation area of 33,800 hm^2.

The water diversion and transmission structures are distributed along a continuous mountainous area crossing towering peaks with a long water transmission line of which the total length of branch canals and above is 880 km. The canal system includes many structures that are mostly made up of tunnel groups.

The general trunk canal and trunk canal project consists of 71 tunnels with a total length of 110 km. It passes through areas with harsh natural conditions, such as large burial depths and soft rocks. The geological condition of the project is extremely complex, with high construction difficulties. The project has 38 aqueducts, among which the high-bent aqueduct at Zhuanglang River in the East Second Trunk Canal is 2,194.8 m long. It has three inverted siphons, among which the Xianming inverted siphon has a 107 m high design head and is 524.8 m long with a double row steel pipe of 2.6 m diameter installed. It ranked first in Asia in the mid-1970s for its scale.

After the project started its construction, it was suspended in 1980 due to the limited construction funds and technical conditions. In 1985, construction resumed. Since its completion in 1995, the project has been in good operation. As of December 2000, the coordinated irrigation area reached 33,300 hm^2 with 42,000 migrants settled, which has achieved considerable economic, social, and environmental benefits.

THE WATER DIVERSION PROJECT IN THE UPPER REACHES OF SHANGHAI HUANGPU RIVER

China's Largest Urban Water Supply Project

The Water Diversion Project in the Upper Reaches of Shanghai Huangpu River is a large-scale urban infrastructure to improve the raw water quality of Shanghai waterworks. It is also the largest urban water supply project in China. The project scale is 5.4 million m³/d in volume and was implemented in two phases. The water intake location is close to Huangpu River's Songpu Bridge, upstream of the Nv'er Jing's[1] outflow, which is a section of relatively better water quality throughout the river. The project started in July 1994 and was completed and launched on December 19, 1997, with a total investment of 2.66 billion yuan.

Riverside Balancing Reservoir in the Pumping Station

---

1. T.N. (Translator's Note): Nv'er Jing is a 4.5 km long river that flows through the southwest area of Minhang District in Shanghai and runs into the Huangpu River.

The main project comprises water intake and booster pumping stations, water transmission channels, large river crossing steel pipes, and corresponding power supply, instrumentation, communication, and dispatching works.

(1) Water intake structures at the bridge. They are composed of four reinforced concrete cylinder water intake structures in the center of the Huangpu River, 8 m in diameter and depth, where four steel pipes with a full length of 140 m and diameter of 3.5 m extended to the pumping station suction well.

(2) Pumping station. It consists of three main pumping stations, Daqiao, Linjiang, and Yanqiao. The main structures of the pumping station include a pump room, a balancing reservoir, a regulating pool, a gate conversion well and a substation, and a distribution room of 35 kV. There are also seven regulating pumping stations for transmitting water into each water plant, with a total design capacity of 3.64 million m³/d.

(3) River crossing pipes. There are three pipes passing through the bottom of the Huangpu River, among which the Linjiang river crossing pipe is 4 m in inner diameter and 720 m in length, with a designed water transmission capacity of 3.1 million m³/d. The Yangpu water plant section crossing pipe is 3 m in inner diameter and 1,200 m in length, including the 600 m long river crossing section. It has a designed water transmission capacity of 1.4 million m³/d. The crossing pipe of the Nanshi water plant section is 3 m in inner diameter and 1,120 m in length, with a designed water transmission capacity of 900,000 m³/d.

(4) Canal. The water transmission trunk line is a cast-in-place reinforced concrete culvert with a total length of about 40 km. There are permeable wells, overflow wells, maintenance wells, and drainage pipes installed along its route.

(5) Power supply and detection and dispatching communication works. The power supply of the three pumping stations is simultaneously supplied through two 35 kV incoming lines with unified dispatching of communication and detection.

During the construction of the Huangpu River water diversion project, a number of new technical measures were carried out, which mainly included the fly ash technology for large volume reinforced concrete, the high power cascade speed control device, synchronous motor out-of-step and then whole step technology for large capacity pumps (single pump capacity 1.6 MW), the underwater jacking technology for large diameter steel pipes, the large diameter rubber live joint technology, the large rectangular cross-section ultrasonic flow measurement technology, the PLC communication scheduling and detection system, etc.

The implementation of the Huangpu River water diversion project has achieved significant social and economic benefits. After the water intake point moved up, the quality of the water source of Shanghai's water supply has been tremendously improved, the majority of which reached the national water source and quality standard Class III.

DAHUOFANG RESERVOIR WATER TRANSMISSION PROJECT

"Lifeline" Project of Liaoning Province

The Dahuofang Reservoir Water Transfer Project includes two phases. Phase I of the project transfers water from Liaodong to Fushun Dahuofang Reservoir. Phase II of the project transfers water from Dahuofang Reservoir to the recipient cities. The water transfer tunnel of Phase I is 853 km in length and 8 m in diameter, which was the longest tunnel under construction in the world at that time. Phase I of the project started construction in March 2003, and Phase II started on September 19, 2006.

Dahuofang Reservoir

Liaoning Province suffers from serious water shortage where most of its regions meet the serious water scarcity standard of the United Nations except for the eastern area. The main goal of the project is to transfer the abundant high-quality water from the mountainous areas of Liaodong to the Dahuofang Reservoir, which will then be delivered to the six cities in central Liaoning, including Shenyang, Fushun, Liaoyang, Anshan, Panjin, and Yingkou to solve the century-long water problem in these areas. The water can also replace groundwater to solve the problem of groundwater overdraft and seawater backflow in central Liaoning. With a beneficiary population of nearly 10 million people, the project is known as the "lifeline" project of Liaoning Province. The water transmission project adopted tunnels and pipelines

Photo of the site of Dahuofang Reservoir Phase I Project

for transmission and delivered a total of almost 1.8 billion m³ of water to 6 cities. Its 231 km long water transmission pipeline crosses seven railroads, 80 roads, 22 rivers, 52 pipelines, and 26 fiber optic cables.

Regardless of its total length of merely 140 m, the six rivers section in phase I of the project is the most complicated section of the entire Dahuofang Reservoir water transmission project due to its burial depth of 63 m and its location at the intersection of three faults with broken rocks and high permeability. In view of the complexity and enormity of the excavation of this cave section, a construction plan of an over-advanced pre-grazing and over-advanced pipe shed was adopted. During the construction, internationally advanced technology was also used for geological forecasting to grasp the geological conditions of the surrounding rocks ahead of the route. A total of 15,487 linear meters of drilling and the sweeping hole were completed, and 4,332 meters of pipe shed were constructed, which solved the construction difficulties caused by broken surrounding rocks and serious water seepage and successfully handled many sudden geological disasters such as water gushing and collapse, creating a miracle in the world of excavating tunnels under complex geological conditions with no accidents.

HUAI RIVER CHANNEL PROJECT
Backbone Project Expanding the Outlet of Huai River Flood

The Huai River Seashore Watercourse Project is located in Huai'an and Yancheng cities of Jiangsu Province, on the north side of the northern Jiangsu main irrigation canal. It is an important flood control project for discharging the flood of the Huai River and Hongze Lake. Based on the existing watercourse, the northern Jiangsu main irrigation canal and the Huai-Yi River Diversion Project can elevate the flood control standard of the lower reaches of the Huai River and the Hongze Lake embankment for a 100-year return period flood in the short term and a 300-year return period flood in the long term.

The Huai'an Hub of Huai River Channel

Hongze Lake is a large lake-type reservoir that can discharge floods from an area of 158,000 km² in the upper and middle reaches. It has a total capacity of 17.6 billion m³, but its downstream flood outlet flow has been historically inhabited. After 1949, with more than 40 years of regulation, its downstream flood discharge capacity has increased from 8,000 m³/s to 13,000–16,000 m³/s. But its flood control standard is still unable to handle a 100-year return period flood. In case of water level rise in both Huai and Yi rivers, the reservoir can only defend against flood with a 50-year return period. Since the 1980s, with the implementation of the river regulation project in the upper and middle reaches of the main stem of the Huai River, the completion of the Cihuai New River Project, and particularly the operation of the Huai-hong-new River Project, the speed of flood flowing into the lake has been accelerated, while the issue of insufficient outlets for downstream flooding stood out more as well. To solve the flood discharge issues of the lower reaches of the Huai River, improve the flood control standard of Hongze Lake, and ensure the safety of the 1.33 million hm² of arable land and nearly 20 million people in the downstream area, additional seashore watercourse projects need to be built.

The west end of the Huai River watercourse project starts from the second estuary of the east side of Hongze Lake. Extending along the north side of the main irrigation canal and the main canal with a "two rivers and three dikes" layout, the project crosses four counties and Huaihai Farm in Huai'an and Yancheng, two cities in Jiangsu Province. It eventually flows into the Yellow Sea north of Biandan Harbour, with a total length of 163.5 km. The project was implemented in two phases: the short-term phase and the long-term phase. The short-term project is designed to discharge 2,270 m³/s of the flood of Hongze Lake, which elevates the flood control standard of Hongze Lake for a 100-year return period flood. It also increases the flood control capacity of the north region of the canal for a 5-year return period flood. The long-term project aims to expand the flood discharge capacity to 7,000 m³/s of the flood, which will elevate the flood control standard of Hongze Lake for a 300-year return period flood and improve the flood control capacity for a 10-year flood for the north region of the canal.

The short-term seashore watercourse project mainly consists of a river embankment, floodway, hubs, cross-river bridges, cross-dike structures and drainage, and irrigation impact treatment project for the north of the canal. The south embankment of the short-term project is heightened and reinforced along the north embankment of the irrigation canal while its north embankment is newly built. The elevation of the embankment is determined by the design flood level plus its super elevation, of which the south embankment is 2.5 m with the superelevation and the north embankment is 2 m with the superelevation. Both have a top width of 8 m. The floodway is divided into north and south floodways. Its scale is designed per the standard of a 5-year return period flood in the northern regions of the canal. Its drainage flow is determined by deducting the volume pumped at the canal north pumping station, and the flow discharged into the main canal and is laid out based on the principle of separate drainage for a flood at different elevations. For the west section of the project, a single floodway was dug in accordance with the earth embankment control. Flood of the beach face being discharged through the south floodway; the East section of the project has two floodways, the south and the north. The waterway intersects with the Second River, the Grand Canal, and the Tongyu River, and eventually flows into the Yellow Sea. Four hubs have been constructed correspondently at the Second River, Huai'an, Binhai, and Haikou, with a Huai-fu control gate built as well.

JINGJIANG RIVER FLOOD DIVERSION PROJECT

Flood Control Project Ensuring the Safety of Jingjiang River Embankment

Jingjiang Flood Diversion Project is located on the south bank of Jingjiang River (right bank) in the territory of Gong'an County, Hubei Province. It stores the flood water that exceeds the safe discharge volume of the Jingjiang River, making it a flood control project that guarantees the safety of the Jingjiang River embankment. The project is also known as the Jingjiang flood diversion area.

The monument of Jingjiang River's flood diversion sluice and inlet sluice

The Yangtze River section between Zhicheng Town in Hubei Province and Chenglingji Section in Hunan Province is 337 km long. It flows through the Jingzhou area in Hubei Province and is called the Jingjiang River. The Jingjiang Dyke is one of the key dykes of the Yangtze River. The dyke is relatively high, with a wide area of land protection. In case of a breach, the vast plain in the north of the Jingjiang River will be inundated by flood, with the possibility of interrupting the navigation of the Yangtze River and threatening the safety of Wuhan City. In order to relieve the flood control burden on the dyke, the Jingjiang River flood diversion area was built in 1952.

The flood diversion area covers an area of 920 km², with a length of about 70 km from north to south and a width of about 30 km from east to west. It is surrounded by dykes on all sides, with an effective flood volume of 5.4 billion m³. The dykes are equipped with temporary housing and warehouses for temporary shelter and the preservation of important materials. It also has a yellow jacket ship lift, more than ten irrigation gates, and three electric drainage pumping stations.

The main project is composed of a cofferdam project, flood diversion gates, flood discharge projects, and flood control gates.

(1) The total length of the dike is 211 km. To prevent wind and wave erosion, willow is planted along the dyke in the area, and the southern dyke adopts dry masonry protection of the slope, while the rest is protected by grass.

(2) Taipingkou Flood Diversion Gate, also known as the North Gate, has a maximum design flow of 8,000 m³/s.

(3) The floodplain drainage project has two operation scenarios ① During the flooding diversion, when the water level reaches 41 m with a 42 m forecast, the flood will be discharged at the artificial levee breach at the Wuliang Temple Dyke; ② When the flooding diversion process is over, and the water level is under 42 m, the flood will first be discharged into the Yangtze River through the artificial levee breach at the Wuliang Temple Dyke once the river level drops and the remaining flood will be released into the Hudu River through the two floodgates on the southern embankment.

(4) Yellow Mountain Head Control Gate is also called the South Gate. When the water level in the flood diversion area reaches 42 m, if a breach happens at the lower section of the Hudu River East Dyke, the water flow of the Hudu River will increase and endanger the safety of the polder on both banks below the head of Yellow Mountain. The flow rate of the control gate will be limited to below 3,800 m³/s.

In 1954, the Yangtze River flooded. The Jingjiang flood diversion project has opened the gates three times and diverted floods for a total of 30 days. During the process, it discharged part of the flood into the Hudu River and Yangtze River, which protected the Jingjiang Dyke from danger. After 1954, the flood diversion area was never used again. With the implementation of measures such as the Jingjiang Dyke reinforcement and cut-off projects, the flood diversion area was used less than when it was first built. With the Three Gorges Reservoir completed preventing floods, the frequency of floodplain usage dropped from once in 10 years to once in 100 years.

TAIHU LAKE REGULATION BACKBONE PROJECT

Regulation and Defense Project of Taihu Lake Basin

Lake Tai Basin Regulation Backbone Project is referred to as the "Lake Tai Regulation Backbone Project." It consists of 11 projects, including Wang Yu River, Tai Pu River, Lakeside Embankment, Hangjia Lake South Drainage, West Lake Tai Diversion Drain, Wu Cheng Xi Diversion Drain, East and West Tiao Xi Flood Control, Lanlu Gang, Hongqi Pond, Hangjia Lake North Drainage, and Main Stem Flood Control in the Upper Reaches of Huangpu River. The project is mainly built for flood control and surface drainage, with comprehensive benefits such as water supply, navigation, and improvement of the water environment potentially offered as well.

Taipu Gate Project

In 1991, after the floods in the Yangtze River and Huai River, the State Council proposed the "Decision on Further Regulation of the Huai River and Lake Tai" to carry out comprehensive water resources regulation of the Lake Tai basin. The planning principle is to take everything into consideration, regulate comprehensively, and develop in all aspects with a phased implementation. Flood control of the basin was designed per the actual rainfall in 1954, during which the maximum 90-day precipitation is equivalent to that of about a 50-year return period flood. Each water conservancy sub-district generally selects its local standard of a 20-year return period short-term rainstorm as the design standard for flood control and discharge. The basin water supply was designed per the actual rainfall in 1971.

The 11 backbone projects for the regulation of Lake Tai can be divided into three categories: Category I projects serve the main objective of guaranteeing safe flood storage and discharge of Lake Tai; Category II projects have the main benefit of regional drainage and water diversion. It also plays an important role in basin flood control. Category III are projects for coordinating flood control at inter-provincial boundaries. Category I projects, which are the backbone projects mainly built to guarantee safe flood storage and discharge in Lake Tai, comprise the following four parts.

(1) Wang Yu River Project. It originates south from the lakeside sand pier of Lake Tai and extends north into the Yangtze River through the Gengjing estuary, with a total length of 60.8 km. The entire project is located in Jiangsu Province and is rated a first-class project. The project mainly includes the river channel project, the Wangting Water Conservancy Hub, and the Changshu Water Conservancy Hub.

(2) Tai Pu River Project. It originates west from Hengshan Street right by Lake Tai, connects with Mao River on the east side, and flows into Huangpu River through Xie Pond. The total length of the project is 57.6 km, of which 40.73 km is in Jiangsu, 1.63 km is in Zhejiang, and 15.24 km is in Shanghai. It is a first-class project that mainly includes the river channel project, the reinforcement of the Taipu Sluice, and the Taipu River pumping station project.

(3) Lakeside Embankment. The total length of the embankment is 282 km, of which 217 km is in Jiangsu and 65 km is in Zhejiang. The project mainly includes embankment earthwork, retaining wall and revetment facing the lake, sluice control structures, and flood control roads along the embankment.

(4) Hangjia Lake South Drainage Project. It is part of the project system of flood discharge into Hangzhou Bay, which is located in the Hangjia Lake region in Zhejiang Province. The project consists of four estuary outlet structures (first-class project) along Hangzhou Bay and the corresponding river channel projects (third-class project). Its Nantai head sluice comes with a supporting river channel of 80 km. The Changshan gate has a 75 km long supporting river channel. Its Yanguan lower river hub includes a tide sluice and pumping station, with a 25 km long supporting river channel. The Yanguan upper river sluice has a 23 km long supporting river channel.

During the development of the Lake Tai Regulation Backbone Project, the project defended three regular floods in 1995, 1996, and 1998, respectively, and the huge flood in the basin in 1999, which created direct economic benefits totaling 15.6 billion yuan in flood prevention and mitigation.

DULIUJIAN RIVER

The Backbone Project Regulating the Hai River Water System

Located in the southwest of Tianjin, the Duliujian River was dug by hand to discharge the flood of the Daqing River system into the sea, ensure the safety of Tianjin and the Tianjin-Pukou Railway, and reduce flooding in the middle and lower reaches of the Daqing River. Duliujian River got its name as it originates from somewhere near Duliu Town.

The floodgate of Duliujian River after risk mitigation and reinforcement

The inlet of the Duliujian River is at the sixth fort of the confluence of the Daqing River and Ziya River. It flows southeast into the Bohai Sea with a design flow of 3,200 m³/s. The river channel is 67 km long and can be divided into three sections from top to bottom: ① Section one is from the intake gate to the upper gate of Beida port. It is 43.5 km long with a designed river bottom longitudinal slope and surface slope both at 1/27,200 degree; ② Section two is the Dagang floodway. It is 17.8 km long with a design surface slope at 1/14,800 degree; ③ Section 3 is the outlet channel east of the port. It is 5.6 km long with a design surface slope of 1/8,100 degrees.

The Beida Port was originally designed for flood control and moist prevention, freshwater storage and irrigation, and water storage and siltation. Later, as the construction of the mine occupied the majority of the Beida Port, it could not fully serve its role anymore. 18.5 km below the floodgate, there are two deep grooves, the north and the south groove, in the river channel along the dike. The deep north groove has a ditch inside for navigation, which was also used for the drainage of the bottom water of the east pond. There are deep grooves on both sides of the Dagang floodway and outlet channels east of the port as well. The north groove of the Dagang floodway also functions as a navigation channel, with a flat in between its two deep grooves for flood discharge. Adjacent to the estuary, Dagang Oilfield and Beida Port Reservoir are located south of the south dike, while Dagang Power Plant, Petrochemical Plant, and the oilfield and natural gas field of Dagang are located north of the north dike.

There are two intake gates at the upper gate of the Duliujian River. The axes of the north and south gates are about 450 m apart. The lower gate of the river is built with a worker-peasant-soldier tide sluice with a designed flow of 3,200 m³/s. Because the main stem of the Hai River cannot reach the original design flow at 1,200 m³/s, in order to ensure flood control safety in Tianjin, the main stem of the Hai River is regulated at 800 m³/s instead, with the remaining 400 m³/s to be discharged by Duliujian River, resulting in an increase in the design flow of Duliujian River from 3,200 m³/s to 3,600 m³/s. Therefore, the worker-peasant-soldier tide sluice was reconstructed in 1994, and the north dike of the Dagang section was elevated and reinforced in 1995 per the standard of 3,600 m³/s. The planning aims to elevate the flood discharge level of the river channel with more measures to be taken to achieve the standard of 3,600 m³/s.

Since its completion, the project has played a great role in flood control in Tianjin. In 1954, the maximum flow rate at the north intake gate was 1,370 m³/s, which was the highest flow rate ever since the construction of the gate. In 1956, the maximum discharge was 1,190 m³/s. In 1963, the maximum discharge of the north intake gate was 1,220 m³/s during the catastrophic flood in the Hai River basin, with 2.66 billion m³ of flood released into the sea through the Duliujian River in total. In the above three high flow years, the north intake gate functioned exceeding its design standard three times and withstood the test of the flood. In 1977, the maximum flow rate of the South Gate was 568 m³/s, which greatly reduced the pressure of the flood on the North Gate.

After the floodgate of Duliu River operated for forty to fifty years, problems such as ground settlement, serious siltation, poor discharge, anti-slip and anti-seepage stability not meeting the design requirements, and aging equipment emerged. On April 6, 2005, the floodgate was reinforced with risk mitigation. The project was completed by the end of 2007, after which the total design flow of the two sluices increased to 3,600 m³/s, enabling the river to play a greater role in flood control, drainage, irrigation, and water supply.

DONGPING LAKE FLOOD DIVERSION PROJECT

The Last Level of Water Storage Lakes of the South-North Water Transfer Eastern Route Project

Dongping Lake flood diversion project is located at the confluence of the Yellow River and Wen River across three counties in Shandong Province, including Dongping, Liangshan, and Wenshang. The project is built to detain and store the floods of the Yellow and Wen Rivers. It controls the discharge flow of the Aishan Station of the Yellow River to ensure the embankment safety of the lower reaches of Jinan City, Jinpu Railway, Shengli Oil Field, and the Yellow River downstream embankment. It is an important part of the flood control system of the lower reaches of the Yellow River and the last level of the reservoir lake along the eastern route of the South-North Water Transfer Project.

Dongping Lake flood diversion sluice

In 1855, after the Yellow River took over the channel of the Daqing River to flow into the sea, Dongping Lake was connected with the Yellow River and became a natural flood detention area of the Yellow River. It was first called the Dongping Lake flood detention area in the Republic of China. After the flood in 1958, the Wei Mountain Water Conservancy Hub was built, which transformed Dongping

Lake into a reservoir with flood control that provides comprehensive benefits like irrigation, power generation, water transportation, and aquaculture. Later, due to the serious siltation of the river channels above the Wei Mountain hub, the hub was destroyed and scrapped in 1963, making Dongping Lake an independent flood diversion project once again.

The flood diversion area is 627 km² in total, with an operating water level of 45 m and a corresponding reservoir capacity of 3.354 billion m³. The project mainly includes the embankment (including the secondary lake dike), the flood diversion sluice, the flood discharge (exit) sluice, etc.

(1) Embankment. It is 100.08 km long, generally 8–10 m high, and 10 m wide at the top. The riverside slope is 1:3 with wave-proof dry masonry for slope protection. The slope against the river is 1:2.5 with a back berm (top width 4 m, side slope 1:5). The total length of the secondary lake embankment is 26.73 km. It was elevated and reinforced per a 48 m crest elevation at the end of the 20th century, which can raise the water storage level of the old lake to 46 m.

(2) Flood diversion sluice. There are three flood diversion sluices in the flood diversion area. Shiwa flood diversion sluice diverts the flood to the new lake, while the Linsin flood diversion sluice and Shilibao flood diversion sluice divert the flood to the old lake. The total design flood flow of the sluice is 8,500 m³/s.

(3) Flood sluice. There are three flood sluices in the flood diversion area: Chenshankou, Qinghemen, and Sigai. The total design flow of the three sluices is 3,500 m³/s.

To successfully discharge and detain the flood of Dongping Lake, the first step is to fully utilize the diversion and storage functions of the old lake before involving the new lake. When the old lake is out of capacity for the flood, the old and new lakes can be used jointly. Due to the siltation in the Yellow River, it is difficult to discharge the north part of the lake into the Yellow River. When the flood flow of Wen River is too large, the second level of the lake dike can also be broken to utilize the entire lake for flood discharge. If there's remaining flood to discharge in the south lake after the entire lake was utilized, approval from the State Council would be needed for discharging flood to the four lakes in the south. Since the completion of the project, it has diverted and stored water from the Yellow River twice (1960 and 1982) as well as utilized the old lake for storing water from the Wen River once every year.

As the South-North Water Transfer Project plans to use the old lake area of Dongping Lake to store water, the old lake will be used at a high water level in the long run, carrying out various functions such as water storage, water supply, water transfer, and irrigation, and playing an important role in alleviating the water resources shortage in the north and the lower reaches of the Yellow River.

PEOPLE'S VICTORY CANAL IRRIGATION DISTRICT

The First Large-Scale Gravity Irrigation District at the Lower Reaches of Its Water Source, the Yellow River, since 1949

People's Victory Canal Irrigation District is located in the northern part of Henan Province. It is the first large-scale irrigation district built in the lower reaches of the Yellow River after the founding of the People's Republic of China, utilizing the water of the Yellow River for irrigation. The total control area of the irrigation district is 1,486 km². It mainly irrigates the 98,000 hm² of arable lands in 47 townships from eight counties, including Xinxiang, Anyang, and Jiaozuo. It also undertakes the urban water supply of Xinxiang City and the water supply to Anyang and Jiaozuo when necessary.

Panorama of water intake port of the People's Victory Canal

The design flow of the project is 60 m³/s, and the increased design flow is 85 m³/s. The project started its construction in 1951, with the first phase of the project completed in April 1952. On October 31, 1952, Chairman Mao Zedong personally inspected the People's Victory Canal and issued the great call to "handle all Yellow River-related affairs well" during the inspection. The head of the canal is located on the Qinguang embankment, 1,500 m west of the Beijing-Guangzhou Railway Yellow River Bridge on the north bank of the Yellow River. The opposite bank is Taohuayu, the dividing line between the middle and lower reaches of the Yellow River. The People's Victory Canal is thus located at the top of the lower reaches of the Yellow River.

The irrigation district has a continental monsoon climate of a warm temperate zone, with an average annual temperature of 14°C. It has a frost-free period of 220 days, an average annual water evaporation of 1,300 mm, and an average annual precipitation of 620 mm. Its soil quality is mostly composed of light loam and medium loam, with wheat, corn, cotton, rice, and peanuts mainly grown in the district. Since the opening of the People's Victory Canal Irrigation District, it has brought great economic, social, and ecological benefits. Before its opening, droughts and floods were frequent in the irrigation district, and the society and economy were very poor and underdeveloped. After its opening, drought, flooding, salinization, and sedimentation have been comprehensively regulated. Grain and cotton yields increased year by year, reaching an average grain yield of 14,250 kg per hectare and a cotton yield of 1,125 kg per hectare by the end of the 20th century, which is 107 times and five times the yield before the opening of irrigation, respectively. The area also uses the sediment from the Yellow River for land reclamation. It transformed the low-lying desolate saline-alkali land into farmland with high stable yields and reclaimed more than 6,000 hm² of land from silt.

In order to solve the salinization of the farmland in the irrigation district, the irrigation district has gradually implemented planned water use, promoted the combination use of surface and groundwater, established a set of water and salt monitoring and water distribution systems, and carried out scientific research on the improvement of saline-alkali land since 1954.

In terms of the siltation and sediment of the diversion of the Yellow River, the area adopts sedimentation ponds for centralized treatment

Channel of the People's Victory Canal

and actively conducts canal system adjustment to improve the sediment transport capacity of the channel for transporting the sediment to the field. The measure not only reduced channel siltation, but also upgraded soil fertility, which set a precedent in diverting the Yellow River for irrigation in its lower reaches. Since the 1990s, the irrigation district has started carrying out more technical transformation projects for water conservation and silt reduction to meet the requirements of social and economic development.

QINGTONGXIA IRRIGATION DISTRICT
A Place in Northwest China with South-China Scenery

Qingtongxia Irrigation District is located in the Ningxia Hui Autonomous Region on the floodplain of the Yellow River. It borders Helan Mountains at the west and the Ordos Tableland at the east. It is composed of eight cities, including Qingtongxia, Wuzhong, Lingwu, Yongning, Yinchuan, Helan, Pingluo, and Shizuishan.

The terrain of Qingtongxia Irrigation District is high in the south and low in the north, with the Yellow River running through. It has abundant water sources, fertile soil, average annual precipitation of about 200 mm, and sufficient sunshine, making it suitable for developing irrigated farming. As early as the Qin and Han dynasties, the locals started opening up canals to divert water to irrigate the farmland for more than 2,000 years now.

After the founding of the People's Republic of China, the irrigation district has undergone large-scale expansion and renovation. The Qingtongxia Water Conservancy Hub Project was completed in the mid-1960s, ending the long history of no dam to divert water in China. The irrigation area is designed

Qingtongxia Water Conservancy Hub panorama

to irrigate an area of 38.8 million hm², with rice and wheat as main crops and some horsetail and sugar beets grown too. It is an important grain-producing area in northwest China and is known as "a place in Northwest China with the scenery of the south Yangtze River."

Outdoor units of Qingtongxia Hydropower Station

The irrigation district can be divided into two major systems, the East and the West. The west general main canal diverts water from the bottom of the dam and divides into another four main canals, the West, Tanglei, Huinong, and Hanyan. The east irrigation district has two main canals, the high canal and the low canal. The high canal diverts water above the dam, while the low canal diverts water below the dam and connects to the Qin and Han canals. Main problems that the irrigation district faces are: ① High irrigation water consumption. Since the late 1990s, measures have been taken, including anti-seepage masonry protection of the main canal, an increase in the water bill, the implementation of planned water use, and strengthened water management, which achieved significant results; ② Yinbei area, which is located in the lower reaches of the irrigation district, has suffered from poor drainage, high groundwater level, secondary salinization that affected its yield. The area also has a large amount of arable wasteland. Thus, there is great potential for water conservation, yield increase, and planting structure transformation in the area. By further improving drainage conditions, implementing scientific water-saving irrigation systems, and accelerating saline-alkali land management, the area can make a greater contribution to the development of the economy and agriculture in the entire Ningxia Hui Autonomous Region.

SHAOSHAN IRRIGATION DISTRICT

Irrigation-Focused Multi-Purpose Project in the Hilly Area
of the Middle Reaches of Xiang River Basin

Shaoshan Irrigation District is located in the hilly area of the middle reaches of the Xiang River basin in Hunan Province. It irrigates 6.67 million hm² of farmland in seven counties, including Xiangxiang, Xiangtan, Ningxiang, Shuangfeng, Shaoshan, Wangcheng, and Yuhu, within a radius of about 2,500 km². It is a large-scale irrigation-focused water conservancy project with comprehensive functions like power generation, shipping, flood control, drainage, and urban water supply available as well.

Shaoshan Irrigation District Reservoir Hub panorama

The construction of the irrigation district started in 1965. It started operation in 1966, and all the supporting projects were completed in 1969. The entire project consists of three parts: reservoir hub, water diversion hub, and irrigation district project.

The reservoir hub is located at the Shuifu Temple in Shuangfeng County, which is in the middle reaches of Lianshui. It is made up of a dam, a power station, and a navigation lock. The control basin area is 3,160 km², with a regular storage level of 94 m and a total reservoir capacity of 0.56 billion m³. The dam is a masonry gravity dam that consists of an overflow section and a non-overflow section, with a maximum height of 35.8 m. The installed capacity of the power station is 4 × 7.5 MW. The navigation lock is a grade II single lock with an annual design freight capacity of 700,000 t.

The diversion hub is located at Yangtan of Xiangxiang City, which is 18 km from the lower reaches of the reservoir hub. It is composed of a barrage, a power station, a canal-inclined plane, and an intake gate. The control basin area above the dam is 5,050 km^2 (including 3,160 km^2 controlled by the reservoir). The regular water diversion level is 665 m high, with a corresponding reservoir capacity of 21 million m^3. The dam is 12 m high, of which the power stations have an installed capacity of 3 × 0.5 MW. The annual design freight capacity of the inclined boat lift is 120,000 t, and the maximum navigable vessel is a wooden ship of 20–30 t.

The irrigation district project has five components: irrigation ditch project, canal structures, irrigation project, flood control and drainage project, and small pond and dam project. The irrigation canals can be divided into the main canal, the south, and the north canal, of which the head of the north canal has a designed inlet flow of 45 m^3/s. The design irrigation guarantee rate is 89%, and the average annual irrigation water supply is 5.50 × 10^8 m^3. There are 26 aqueducts, ten tunnels with a total length of 12.5 km, and more than 2,300 canal structures of various sizes on the canals.

After the completion of the irrigation district, millions of acres of farmlands along the canal have been protected from droughts and floods. Compared with the data before the project completion, the planting area of the double-cropping rice has expanded from 25% to more than 95%, and the grain yield per hectare has increased from 3,600 kg to more than 15,000 kg. The 10,000 hm^2 of hilly land along the canal has been developed, and the 10,000 hm^2 of food and cash crops along the Lianshui River have been protected from flooding. The district also supplies water of 45 million m^3 to more than 30

Irrigation ditches of the Shaoshan Irrigation District

industrial and mining enterprises for production and living every year, which reduces the production cost of enterprises by 13 million yuan. 8,700 hm^2 of water surface in the irrigation area has been used for fish farming. Twenty-two small and medium-sized hydropower stations have been built using the reservoirs and channels of the irrigation district, with an annual power generation capacity of about 1.4 billion kW·h. The yield of agricultural products such as fruits, pigs, Xiang lotus, and tea in the benefited area has increased substantially. The green coverage rate of the channel reached 100%, forming a three-dimensional green structure combining trees, shrubs, and grasses with remarkable ecological benefits.

PISHIHANG IRRIGATION DISTRICT

China's Largest Irrigation Area in the Hilly Region

Located in the hilly area between the Yangtze River and Huai River in central and west Anhui, the Pishihang Irrigation District is the general term for the three adjacent irrigation districts at Pi River, Shi River, and Hangbu River. It is a comprehensive utilization project that focuses on irrigation and also supports functions including power generation, shipping, aquaculture, and urban and rural water supply.

The Pishihang Irrigation District covers both Lu'an and Hefei cities with a total area of 13,130 km², of which the hilly area accounts for 84%, and the plain area accounts for 16%. The average annual precipitation in the irrigation area is between 1,006 mm and 1,104 mm, and the average annual evaporation is between 1,009 mm and 1,113 mm. The average annual temperature

Pishihang Irrigation District panorama

is between 14.9°C and 15.7°C with a frost-free period between 215 and 230 days. The main crops of the district are wheat and rice. The irrigation district started construction in 1958 and was put into operation in 1959. Since then, it has been optimized year by year with a designed irrigation area of 0.733 million hm², of which 80% is self-flowing irrigation and 20% is lifted irrigation. By 2000, the effective irrigation area has reached 0.683 million hm². There are 349 backbone canals (general main canal, main canal, sub-main canal, and branch canal) available in the irrigation area, with a total length of 4,730 km. The water utilization coefficient of the canal system was 0.5 in 1999. The irrigation area and the design diversion flow of the irrigation ditches are shown in the table below.

Irrigation Area and Irrigation Ditch Design Diversion Flow of the Irrigation District			
Irrigation District	Design Irrigation Area (million hm²)	Effective Irrigation Area (million hm²)	Design Diversion Flow (m³/s)
Pi River Irrigation District	44	42	330
Shi River Irrigation District	19	18.2	145
Hangbu River Irrigation District	10.3	9.1	105

The major water supply projects in the irrigation districts are five large reservoirs, including Foziling, Mozitan, and Xianghongdian on the Pi River, Meishan on the Shi River, and Longhekou on the Hangbu River. The total design capacity of the five reservoirs is 6.593 billion m³. The capacity of the balancing reservoir is 2.07 billion m³ in the flood season and 3.015 billion m³ in other seasons. There are also 24 medium-sized reservoirs,

View in the Irrigation District

1,112 small reservoirs, and more than 210,000 small reservoirs in the irrigation district, with a total storage capacity of 1.96 billion m³. These projects not only have a counter-regulatory impact on the large reservoirs in the upper reaches, but also adjust and store local runoff and recharge the water supply of the irrigation district. In addition, there are more than 470 pumping stations in the irrigation district with a total installed capacity of 75 MW, of which 39 are water charging stations that can extract more than 100 million m³ of water from rivers and lakes every year to make up for the shortage of canal water.

The irrigation district has a management committee, whose permanent establishment is the General Administration of Pishihang Irrigation District. Each county within the irrigation district has a management office respectively, which is responsible for regulating irrigation operations within its own territory.

Mozitan Reservoir, one of the dams of the water source reservoirs of the irrigation district

Foziling Reservoir, one of the dams of the water source reservoir of the irrigation district

THE EMBANKMENT OF YELLOW RIVER (LINHUANG DIKE[1])

Main Components of the Flood Control Project System in the Lower Reaches of the Yellow River

The embankment projects in the lower reaches of the Yellow River first formed in the middle stage of the Spring and Autumn Period of ancient China and developed on a considerable scale around the Warring States period. Nowadays, there are all kinds of embankments in the lower reaches of the Yellow River, with a length of 2,291 km. The Linhuang Dike, which is the so-called Embankment of the Yellow River, is 1,371.22 km long.

Embankment of the Yellow River

The Linhuang Dike consists of the left and right banks. It is an important component of the flood control engineering system in the lower reaches of the Yellow River, with a protected area of about 120,000 km².

---

1. T.N.: "Linhuang" means in adjacent to the Yellow River in Chinese.

(1) The right bank of the Linhuang Dike is 624.24 km long. It can be divided into four sections from top to bottom: ① Mengjin dike, which starts from Niuzhuang in Mengjin and ends at Hejia Temple with a total length of 7.60 km. ② The second section starts from the foot of Mangshan in Zhengzhou, Henan Province. It extends past Zhongmou, Kaifeng, and Lankao in Henan Province, as well as Dongming, Heze, Jancheng, and Yuncheng in Shandong Province. This section is 340.18 km long. ③ The third section includes ten dikes which start from the Liangshan area and end at the Qinglongshan area of Dongping Lake with a total length of 19.32 km. ④ The fourth section starts from Songjiazhuang in the suburb of Jinan City. It goes through Licheng, Zhangqiu, Zouping, Gaoqing, and Boxing and ends at the 21 households in Kenli District with a total length of 257.14 km.

(2) The left bank of the Linhuang Dike is 746.979 km long, which can be divided into five sections from top to bottom: ① The first section starts from Zhongcaopo in Mengzhou City of Henan Province. It goes past Wenxian, Wuzhi, and Yuanyang and reaches E'wan in Fengqiu City with a length of 171.05 km. ② Guanmeng Dike, which starts from E'wan in Fengqiu City and ends at Wu Tang with a length of 9.32 km. ③ Taihang Dike, which starts from Dacheji in Changyuan City and ends at Sudongzhuang with a total length of 22 km. ④ The fourth section starts from Dacheji in Changyuan City of Henan Province, passing through Puyang and Fan county, and ends at Zhangzhuang of Taiqian county with a total length of 194.48 km. ⑤ The fifth section includes four parts starting from Tao City in Yanggu County in Shandong Province. It extends via Dong'a, Qihe, Jiyang, Huimin, and Binzhou and ends at Lijin with a total length of 350.12 km.

The dike has gone through generations of elevation and repair. Due to the long dike line, frequent repair, and mass construction conducted by local residents, serious seepage or soil flow occurred during the floods. The main measures that have been taken to reinforce the embankment are cone drilling pressure grouting, slotting and soil replacement, sloping core and blanket with clay, inverted filter drainage, berm built at front and rear, and embankment reinforcement through desilting.

Since the 1970s, embankment reinforcement through desilting has been extensively promoted, utilizing the high sediment characteristics of the Yellow River water. By taking desilting measures such as self-flow, water lifting, simple sand dredger, and mud pump, a total length of 755.6 km of alignment was opened up in the siltation area within 30 years, which played a significant role in consolidating the embankment in the lower reaches of the Yellow River. After reinforcement, potential hazards and weaknesses still exist in the embankment. For example, some sections of the embankment body cannot meet the requirements of seepage stability as per the design flood level of a 200-year return period flood. The majority of the foundation of the buttress dam in critical levee sections is relatively fragile. The slope of the stone base is too steep with low depth, which leads to poor stability. In case of greater floods, the possibility of an embankment breach still exists. Due to the high siltation in the riverbed, the crest elevation of some parts of the embankment cannot meet the requirements of the design flood of a 2,000-year return period flood.

Since 2002, the Yellow River Water Conservancy Commission of the Ministry of Water Resources has proposed standardized embankments construction in the lower reaches of the Yellow River to build a "three-in-one" standardized embankment system of flood protection line, emergency rescue traffic line, and ecological landscape line to seek long-term peace and stability in the lower reaches of the Yellow River. Through the construction of projects such as embankment reinforcement through desilting, embankment widening, embankment top hardening, and wave prevention forests to build the infrastructure that maintains the well-being of the Yellow River and achieves harmony between human and natural environment.

BEIJIANG[1] DIKE

A Solid Flood Barrier in the Pearl River Delta

Beijiang Dike is located on the left bank of the lower reaches of the Bei River. It is an important barrier to defend Guangzhou against floods from the Xi River and Bei River, making it a National Level I dike. The north end of the dike starts from Qibeiling in Shijiao Town in the Qingcheng District of Qingyuan City. It flows south out of the main stem of the Bei River along the Dayan River and goes past Datang, Lupu, Huangtang, Hekou, and Xinan Town in the Sanshui District. The dike ends at the Lion Mountain at Xiaotang Town in the South China Sea District, with a total length of 63.35 km. Along its route, there are 29 sluices across the dike, including two large-scale flood diversion sluices, Lubao and Xinan.

Beijiang Dike (Huangtang Dike section) with reinforcement up to standard

The Beijiang Dike Protected Area has been frequently and heavily flooded historically. After the founding of the People's Republic of China, all dikes that were originally independent and scattered have been connected along the left bank of the Bei River for repair. In 1957, the Lubao sluice was repaired, and the Xinan sluice was built, which formed the comprehensive flood protection system of the Beijing Dike.

---

1. T.N.: "Beijiang" means north of the Pearl River in Chinese.

From 1970 to 1971, the second repair of the Beijiang Dike was conducted by the locals. After the heavy flood in 1982, it was identified that the dike only met the flood control standard of a 20-year return period flood, but not the 100-year return period flood for Guangzhou. Therefore, the third large-scale repair and reinforcement of the dike were carried out between 1983 and 1987 per the flood control standard of a 100-year return period flood. In 1994, the dike successfully defended a 50-year return period flood.

The repair of Beijiang Dike in 1954

After three large-scale repairs and reinforcement, the flood control capability of Beijiang Dike was improved. However, due to the limitation of social and economic conditions at that time as well as historical reasons for the dike itself, the following problems still remain for the dike: serious seepage in the permeable dike foundation and low-quality soil filling in the old dike body; riverbed incision with water flowing to the dike top; thalweg approaching the dike and even scouring at the foot of the dike with an increasing number of critical dike sections; dike structures have been under operation over 40 years with old and aging equipment in disrepair, which seriously threatens the flood control safety of the dike; the dike section and its flood protection capability cannot meet the requirements listed in *Code for Design of Levee Project (GB 50286-98)*. The grade of the dike and the corresponding structures need to be elevated. The slope of the dike needs to be slowed down, the safety superelevation and the width of the dike top need to be increased, and the flood control road and other basic equipment need to be improved. In addition, the two sub-floodways, Lupao and Xinan, are in disrepair with significant potential hazards and critical sections. The siltation of the river is too high to meet the original design flood diversion requirements. The water environment has been continuously deteriorating, which has threatened the health of many residents along the route.

On October 29, 2003, the Beijiang Dike Reinforcement and Compliance Project was approved by the State Council and officially started. The projects reinforced the Beijiang Dike per the flood control standard of the national Level 1 embankment and a 100-year return period flood. The entire project was completed by the end of 2007. Complied with the standard, the Beijiang Dike has become not only a real solid flood barrier in the Pearl River Delta but also a green corridor beautifying the cityscape along the route with its 63 km long main dike.

The West Embankment in Lianyungang
China's Longest Breakwater

The West Embankment in Lianyungang is located in Lianyun District, 30 km southeast of the city center of Lianyungang, Jiangsu Province. It stretches over the Lianyun Strait (also known as Yingyou Strait), an isthmus between the east and west of Lian Island and Yuntai Mountain, extending from Huangyingzui in Xugou to west Lian Island on the opposite side. It is a key project that started during the "Seventh Five-Year Plan" and is planned to be completed during the "Eighth Five-Year Plan." It was known as the "Great Wall of the Sea" and the "First Dike of China." The main goal of the project is to protect the port from wind and waves as well as develop the port. It also contributes to comprehensive benefits such as tourism development.

The West Embankment in Lianyungang

The construction of the embankment started on March 10, 1985, and was successfully completed on December 8, 1993. It's been more than 20 years since its first operation, during which it provided an excellent natural barrier for the port area of Lianyungang. It connects Lian Island to the inland, reclaims numerous lands through silt flushing, and lays a solid and powerful foundation for the development of the port.

The West Embankment of Lianyungang is built on the underwater shallow on the north side of Xugou Bay in Lianyungang, of which the average water depth is –1 to –2 m. It is 6.7 km long in total and is the longest rockfill breakwater in China. The top width of the dike is 12 m. The net width of the road is 10 m. The height of the embankment body is 7 m above sea level. The top elevation is over 9 m high, with a 7.8 m high curved seawall built on the embankment. The main purpose of the embankment is to block the wind and waves from the northwest, relieve the jacking force from the east and west sides of the channel at high tide, and reduce the siltation of sediment.

After its completion, the area of the water surface of Lianyungang increased from the original 3 km² to 30 km², adding more than 8,000 m of shoreline to the port of Lianyungang, which has been lacking shoreline. The area has the capacity to build more than 40 berths for million tons, making the Port of Lianyungang a semi-enclosed superport.

Due to the special geographical location of Lianyungang, silt thicker than 10 meters exist under the port area, making it extremely difficult to build the breakwater. In order to develop the Lianyungang Port better, establish its status as the Great Oriental Port, and develop its island resources, many experts and scholars have conducted nearly 20 years of scientific research, investigation, and testing, overcoming many difficulties. As a result, the construction of the West Embankment was approved by the State Council. Thanks to its successful construction, Lianyungang has become an important stop for joint transportation by land and sea on the "New Silk Road," presenting itself on the stage of the world economy with a new look.

The West Embankment was built not only to better exploit the island resources and control the sedimentation, but also to improve the ecological environment of the surrounding areas, develop diverse tourism and advance economic development. Between 2014 and 2015, the West Embankment was renovated. Its lane was expanded to a 28.25 m wide four-lane road in both directions, and the road surface was changed into asphalt concrete pavement with infrastructure like green belts and sidewalks added. It has improved the travel conditions for the locals and created a good tourist environment for visitors. The night scene of Port of Lianyungang is magnificent, with lights more than 10 km long along the coast. The west embankment looks like a golden dragon crossing the sea, with glittering signal lights and dazzling beacon lights around. A transparent crystal palace has shown up with the slow-moving graceful lights from the ships in the sea, the on-and-off lights on the fishing boats, and the lights from the households on the opposite shore.

Sichuan River Waterway

The Main Waterway Connecting Southwest China to East China

The 1,045 km-long waterway, which starts from Yibin City, Sichuan Province, to Yichang City, Hubei Province, in the upper reaches of the Yangtze River, is commonly known as the "Sichuan River." Sichuan River, together with the Jinsha River, Min River, Tuo River, Jialing River, Chishui River, and Wu River, has formed the water transport network of southwest China, becoming the main water transport channel from southwest China to central and eastern China and coastal areas.

Sichuan River Waterway

The Sichuan River Waterway has always been known for being bendy, narrow, shallow, and dangerous. The 385 km long river section from Yibin to Chongqing is located in a hilly area, with a wide valley and similar width of river sections. The flow velocity on the shoal surface is generally about 3 m/s during the dry spell. The flow is usually in a good pattern except for a few shoals.

The lower section of the Sichuan River (Three Gorges Waterway) is the 600 km river section from Chongqing to Yichang, of which the water head is 125 m. There are 139 shoals in the section, threatening the safety of navigation on the Three Gorges of the Yangtze River.

After 1949, China began to develop the Sichuan River Waterway, with a total of 123 shoals of various types regulated. After the Gezhouba Water Conservancy Project was closed for operation in 1981, the

The "guiding light" on the Yangtze River Waterway

waterway between its permanent backwater region and Xiangxi, which is about 70 km long, was improved. After the comprehensive regulation, the scaled water depth of the waterway between Yichang and Chongqing increased from 2.9 to 3.2 m. The waterway is 60 m wide, with a radius of curvature of 750 m. During the dry spell, a large passenger freighter fleet can navigate throughout the day. The scaled water depth of the waterway between Chongqing and Yibin increased to 2.7 m, with a width of 50 m and a radius of curvature of 560 m, which supports the navigation of a barge fleet of 1,000 t during the dry spell.

The regulation of the Sichuan River Waterway is a successful example of the regulation of river sections in large mountainous regions in China. It has greatly improved the natural barrier of the waterway. Because dangerous shoals obstruct navigation for different reasons, the regulation approach varies as well. For rapids, clearance and explosion at the shoal navigation channel are mostly adopted to expand the cross-section of water flow with a dam built in its lower reaches to bank water up and reduce the gradient and flow velocity of the shoal. The shape of rapids can also be utilized for regulation. The aligned river mouths were changed into offset, which allows ships to use the subcritical flow area above and below the convex river mouth to move from one side of the bank to the other to cross the shoal. For bendy shoals with banks of alternating shapes, depending on the width of the river surface, the construction of a submerged dike on the upper part of the deep groove of the concave bank is a major regulation approach adopted. In this way, the depth of the deep groove and unit discharge decreases. Another approach is to build spur dikes on the upper reaches of concave banks and separates the main stem, which can regulate shoal and shallow at the same time. For shallows, a combination approach of regulation and dredging is normally adopted, which builds spur and longitudinal dikes to narrow down the river width, concentrate and guide the water flow to flush the waterway and meet the requirements of design navigation depth. In the case of shallow beds with tight pebble structures, dredging should be considered too.

After the completion of the world-renowned Yangtze River Three Gorges Project, the backwater in the reservoir area can reach more than 60 km beyond Chongqing City. The 680 km long navigation channel of the Sichuan River has also been improved. The test results of the mathematical and physical models show that a 10,000-ton fleet can reach Wuhan and Chongqing directly for half a year, and the downstream shipping capacity can reach 50 million tons. However, problems, including the impact of sedimentation in the dam area, the fluctuating backwater area, and the banked-up water area of the reservoir tributaries on the port and waterway, still exist.

A cargo ship loaded with containers passing through the Three Gorges Channel

PORT OF GUANGZHOU

The Comprehensive Main Hub Port in South China

Located in the suburbs of Guangzhou in the Pearl River Delta, the Port of Guangzhou is a comprehensive main hub port in South China.

As early as the Qin and Han dynasties, the Port of Guangzhou has already become the gateway of southern China to foreign trade. For more than 1,000 years, from the Tang Dynasty to the Qing Dynasty, the port was the first point of China's foreign trade, except for the Southern Song Dynasty and the Yuan Dynasty. Before the Opium War broke out, the Port of Guangzhou was the only port in China opening to the rest of the world. It became one of the five treaty ports in 1842. After the Xinhai Revolution, in the *International Development of China*, Dr. Sun Yat-sen put forward the idea of building a large port in south China at Huangpu, which is at the lower reaches of the old Port of Guangzhou. In 1937, a 400 m-long parallel wharf was built along the shore, and a deep channel was dredged. In 1948, the second phase of the project was completed with a 1,250 m long wharf and warehouses, making it a port with a berth capacity for sea vessels of up to 10,000 tons. After the founding of the People's Republic of China, port authority was set up respectively for the two ports at Guangzhou and Huangpu, with a large number of berths being renovated or built. In December 1987, the two ports were merged to form the new Port of Guangzhou. In

Port of Guangzhou panorama

1999, the cargo throughput of the port exceeded 100 million tons, making it the second port in mainland China that became a port of million-ton class internationally. After that, the development of the port accelerated every single year, with the throughput reaching 300 million tons and the container volume of the port exceeding 6.65 million TEU in 2006. Its cargo throughput ranks 3rd among all coastal ports in China and 5th among the top ten ports in the world.

The Port of Guangzhou is located in the waterway networks of the Pearl River Delta, which has a flat terrain. The waterway from the anchorage of Guishan Island at the Pearl River estuary to Huangpu Port is 114.5 km long, with a water depth between 8.6 and 9 m, which allows 25,000 t ships to pass through the Huangpu Port by the tide. The Guangzhou Port Group currently owns 46 berths, 13 cargo handling buoys, and 23 cargo handling anchorages (maximum unit anchoring capacity is 300,000 t) for ships over 10,000 t.

The Port of Guangzhou oversees four major port areas, including the Inner Harbor Area, Huangpu, Port of Xinsha, and Port of Nansha.

PORT OF DALIAN
The Largest Comprehensive Port in Northeast China

Port of Dalian is located in Dalian Bay at the southern end of Liaodong Peninsula, bordered by the Yellow Sea to the east and the Bohai Sea to the west, with Penglai City in Shandong Province can be seen across the Bohai Strait. Port of Dalian is the largest comprehensive seaport in northeast China, known as the "Pearl of the North."

The Port of Dalian has a mild climate with no freeze during the winter in the port area, making it a natural port suitable for navigation in all seasons in northern China. It has a vast port area with a land area of 8 km² and a water area of 346 km². The average water depth in the area is 10 m, and the maximum depth is 33 m. There are no major rivers flowing into the sea along the bay and no impact from longshore drift in the vicinity, leaving the port free from siltation.

Before the 19th century, the Port of Dalian was an important port for China's international transport. After Tsarist Russia leased Lv'da City with force in 1898, the Port of Dalian was declared a free port in 1899 and developed as a commercial port. In 1905, after the Japanese occupied Dalian, the port

Port of Dalian panorama

continued to expand. In 1945, the Soviet Red Army took over Port of Dalian and handed it over to China in 1951. After expansion and new construction, by the end of 1999, Port of Dalian has 73 berths for cargo handling, including 39 berths for ships over 10,000 t, with the largest berth at 100,000 t level, as well as another three buoy berths at 10,000 t level. It has a 146.1 km long railway special line and an 88.6 km long oil pipeline. The public cargo handling wharves of the Port of Dalian distribute across all of its eight-port areas.

(1) Si'ergou Port Area. It has four trestle bend wharves and six berths for cargo handling, including four berths of over 10,000 t.

(2) Dagang Port Area. It has four jetty-type wharves, three parallel wharves, and 22 berths for cargo handling, including nine berths at 10,000 t level.

(3) Xianglujiao Port Area. It has two jetty-type wharves and eight berths for cargo handling, including two berths of over 10,000 t and six berths at 5,000 to 7,000 t level.

(4) Ganjingzi Port Area. It has one steel trestle bend wharf and three berths for cargo handling, including two berths of over 10,000 t.

(5) Nianyu Bay Port Area. It has one trestle bend crude oil wharf and two wharves for oil products. The crude oil wharf has two berths for docking tankers between 30,000 and 150,000 t. The oil product wharf has four berths for docking tankers between 3,000 and 50,000 t.

(6) Dalian Bay Port Area. It has five deep-water berths, which can dock three vessels of 30,000 t at the same time.

(7) Dayaowan Port Area. It has eight deep-water berths, of which five are containers only, and can dock container ships of the fourth and fifth generation. The port area is equipped with the world's most advanced terminal operating system of CITOS1 computer introduced from Singapore. The system has the capacity to process three million TEU per year. There are also two berths for breakbulk cargo at 25,000 t level and one berth for bulk grain at 80,000 t level in the port area.

(8) Heizuizi Port Area. It has eight berths at 6,000 t level.

PORT OF TIANJIN

The Largest Artificial Harbor in China

Port of Tianjin is located at the estuary of the Hai River on the west coast of Bohai Bay, which is also the intersection of the Beijing-Tianjin megapolis and the Circum-Bohai Sea Economic Circle. It is the largest artificial seaport in China and an important international trade port in North China.

The history of the Port of Tianjin can be traced back to the Han Dynasty, with the seaport first formed in the Tang Dynasty. It was officially opened to the public in 1860 and was one of the earliest ports opened for foreign trade in China. The new Port of Tanggu was first built in 1939. After three years of restoration after 1949, it was reopened for navigation on October 17, 1952.

The Port of Tianjin is the largest artificial port in China, which was built through artificial seabed dredging and land reclamation on a muddy shallow. The existing area of water and land in the port is

Port of Tianjin panorama

nearly 200 km², among which land area accounts for 47 km². The total land area of the port is planned to reach 100 km² by 2010. Currently, the main navigation channel is 44 km long, with the maximum width of the channel bottom at 260 m. The maximum water depth of the channel reached 19.5 m, which allows ships of 200,000 t to enter the port freely and ships of 250,000 t to enter by the tide.

The Port of Tianjin has four major port areas, namely, Beijiang, Nanjiang, Dongjiang, and Haihe. It owns more than 140 berths of various types with an 18.1 km long shoreline. There are 78 berths for cargo handling, with a design capacity of 213.99 million t and 5.25 million TEU for containers. The Beijing Port focuses on handling containers and cargo, while the Nanjiang Port is mainly for handling dry and liquid bulk orders. The Haihe Port Area has 15 berths below 6,000 t level, and the Dongjiang Port Area is a new port area with a planned area of 30 km².

Port of Tianjin has a vast economic hinterland with huge potential for development. At present, the service of Port of Tianjin can serve 14 provinces, cities, and autonomous regions, including Beijing, Tianjin, Hebei, and central and western regions in China, with a total area of nearly 5 million km², which accounts for 52% of the national territory. About 70% of its cargo throughput and more than 50% of its import and export value is made up of cargo from provinces and regions outside Tianjin, illustrating its strong radiation and influence on the hinterland. With the revitalization of the economy of Cerium Bohai Rim, the rise of central China, and the advancement of western expansion, there was huge potential for the hinterland economy at the Port of Tianjin, which provided it good conditions for development. In 2005, the cargo throughput of the Port of Tianjin reached 2.4 billion tons, ranking 6th among ports around the world. The container throughput was 4.8 million TEU, ranking 16th in the world container port.

The modernization level of the Port of Tianjin is leading nationwide. By adopting new technologies, equipment and techniques extensively, the Port of Tianjin ranks top among the ports in China in terms of modernization and informatization. It has the most advanced continuous handling facilities in the world.

The coal transport technology through a 10 km-long belt corridor adopted by Nanjiang Port is unprecedented in the world. The Port of Tianjin has also established an international trade and shipping service center and electronic port, which can provide its customers with fast and efficient port "one-stop" port service.

WUHAN NEW PORT
The Shipping Center of the Middle Reaches of the Yangtze River

Wuhan New Port was established in 2010 and consisted of the former Wuhan Port and some other port areas in Huanggang City, E'zhou City, and Xianning City. The total length of the shoreline of Wuhan New Port is 862 km, of which 749 km is the Yangtze River and 113 km is the Han River. The planned port shoreline is 309 km long, of which 76 km is already in use, 112 km is planned to be used, and the rest 121 km is planned for reservation. A total of 1,126 berths for various types of cargo handling are planned, of which 772 are built, 62 are under construction, 74 are for containers, and 13 are for ro-ro (roll-on/roll-off ships). The berths offer a throughput capacity of 617 million tons, including 14.5 million TEU of containers and 1.94 million vehicles of ro-ros.

Night view of Wuhan New Port

Based on the construction of 26 port areas, the plan of Wuhan New Port focuses on developing two major container port areas, one new port business district, five lingang[1] new cities, and twelve lingang industrial parks. The goal is to develop Wuhan New Port into a modern, international hub port that integrates modern shipping logistics, comprehensive bonded services, and lingang industrial development.

(1) The new port business district. The business district will support functions including corporate headquarters, shipping services, commercial trade, corporate consulting, commercial insurance, crew club, tourism, leisure, real estate development, etc. The goal is to build a comprehensive service center for Wuhan New Port and the most important lingang business district in the central region of China.

(2) Two large container port areas, which are Yangluo and Baihshan Port Areas. The development model combines a container port, comprehensive bonded area, and port industrial park, enabling coordinated development between the district and the port.

- Yangluo Port Area: Yangluo Port is the only natural deep-water port in the middle reaches of the Yangtze River. After the completion of the second phase of the project, it focuses on the transport of containers, steel, and breakbulk cargo and mainly serves regions including the north Hubei, south Henan, and the Ankang region of Shaanxi, etc. It allows 5,000 t-class ships to navigate all year round and has an annual throughput capacity of up to one million TEU. Through the port, cargo can arrive in Japan, Southeast Asia, and other offshore countries and regions without transfer, making it the largest container terminal in the middle and upper reaches of the Yangtze River.
- Baihu Mountain Port: The port focuses on the transport of containers, chemical materials, and products. It mainly serves regions in the north and south of Hubei.

(3) Five lingang new port city, which includes Yangluo Lingang New City, Huashan-Beihu Lingang New City, Gehua Lingang New City, Huangzhou Lingang New City, and Xianjia Lingang New City.

- Yangluo Lingang New City: The plan of the city aims to form a modern new city of port industry with comprehensive logistics support, steel and deep processing, grain logistics, and coal transit transportation as its leading industries.
- Huashan-Beihu Lingang New City: The city aims to develop both the Beihu Chemical Industry New City with ethylene and its downstream products, salt and fine chemicals processing, and comprehensive logistics support as its leading industries and a scientific ecological new city that cultivates ecological industries such as science and technology services, leisure and vacation, and tourism services.
- Gehua Lingang New City: The city will become an important scientific and ecological new city which is the modern industrial base for leading industries like electronic information, automotive electronics, biopharmaceuticals, fine chemicals, etc.

---

1. T.N.: "Lingang" means in adjacent to the port in Chinese.

- Huangzhou Lingang New City: The city will become a modern port industrial city with leading industries, including port logistics, deep steel processing, auto parts, new materials and new packaging, light industry and textile, and agricultural products processing.
- Xianjia Lingang New City: The city will turn into a modern lingang industrial new city with a port logistics industry, textiles and garments, forest industry, mechanical and electrical manufacturing, and biomedicine as its leading industries.

(4) 12 lingang industrial parks. The 12 industrial parks will be planned and developed along the port shoreline, which comprises Chibi, Panjiawan, Shamao, Jinkou, Junshan, Hankoubei, Baihushan, Gulong, Tuanfeng, Sanjiang, Zhangjiawan, and Wuzhang-Yangye.

PORT OF NANJING

The Largest Port for China's Inland Navigation

The Port of Nanjing is located in Nanjing, Jiangsu Province, which is a city along the lower reaches of the Yangtze River and is 347 km away from Wusongkou. It is the largest port for China's inland navigation.

The Port of Nanjing has a long history. As early as the Three Kingdoms, it has become a military and commercial port. During the Yuan and Ming Dynasties, it was one of the ports of loading for transporting grains from the south to the north. It was also the base port for the expedition of Zheng He, who was a voyager in the Ming Dynasty. In 1882, the first pontoon-style pier in Nanjing was built. After the founding of the People's Republic of China, the Port of Nanjing has been under large-scale renovation and expansion. Since the Reform and Opening Up era, the Port of Nanjing has achieved particularly rapid development. In 1978, its oil port was opened, making it the largest inland oil port in China. In 1984,

Port of Nanjing

the Xinshengxu Foreign Trade Port was built, making the Port of Nanjing the largest foreign trade inland port of China, In 1987, a Sino-US joint venture, Nanjing International Container Handling Co., was founded, and the Port of Nanjing has thus become the container port with the highest specialization level among all inland ports in China.

In 1990, Nanjing Port Huining Wharfs Co., Ltd. was established, making the Port of Nanjing the most advanced specialized bulk cargo handling inland port in China. At the end of 2002, the ro-ro berth for commercial vehicles was officially put into operation. The Port of Nanjing has thus become the only port with specialized ro-ro berths on the Yangtze River. At the end of March 2004, the Longtan Container Port Area was put into trial operation, which added 520,000 TEU of throughput capacity and gained a higher status for the Port of Nanjing among inland ports. The Port of Nanjing has become a multi-functional port in East China and the Yangtze River basin that supports cargo switch between river and sea transport, land and water transit, cargo distribution, and international exchange. In 2005, its cargo throughput exceeded 100 million tons.

The Port of Nanjing has deep waterways and wide port areas with good natural conditions. The navigation channel from Longzhuayan to Yanziji has a depth maintained of 10.5 m and a width of more than 200 m. The radius of the turn of the channel is much longer than even five times the length of a navigable ship. Sea vessels with a full-load draft of 9.7 m or less can navigate through the port all year round. The public cargo handling terminals of the port are mainly located in Shangyuanmen Port Area, Pukou Port Area, Xinshengyu Port Area, Qixia Port Area, and Yizheng Port Area.

The economic hinterland of Port of Nanjing expands deep into the 12 provinces within the Yangtze River water system. Three main railways intersect the city, including Tianjin-Pukou, Shanghai-Nanjing, and Nanjing-Tongling. Highways that pass the city include Nanjing-Shanghai, Nanjing-Hefei, Nanjing-Yangzhou, Nanjing-Hangzhou, and Nanjing-Wuhu. The waterway of the city goes down the Yangtze River towards the east and reaches Wusongkou after 347 km. In addition to the main tributaries of the Yangtze River, it also connects to the Grand Canal in the east, the water system of the Huai River in the north, and Lake Tai in the south. The port area of Port of Nanjing is also the terminal of the Shandong-Nanjing oil pipeline.

PORT OF SHANGHAI

The Largest Seaport in the World

Port of Shanghai is located in the middle of China's mainland coastline at the front edge of the Yangtze River Delta. It is located at the intersection of the east-west transport channel of the Yangtze River and the north-south transport channel over the sea. It is the world's largest port.

As early as AD 746 in the Tang Dynasty, people developed a port in Qinglong Town (now Qingpu District, Shanghai) for ships to navigate and dock. In the Song Dynasty, Qinglong Town was called "the top trade port in Jiangnan (regions south of the Yangtze River)." Later, due to the siltation of the river channel, the port was relocated to Shanghai in about 1265 with the Maritime Trade Offices set up. During 1403 and 1404, the Huangpu River was formed after regulation on a large scale. After the Opium War in 1840, the Port of Shanghai was forced to open to the public. Since 1853, Shanghai overtook Guangzhou as the largest foreign trade port in China and became the national shipping center after the 1870s.

Port of Shanghai

At the beginning of the 20th century, Huangpu River Channel Bureau carried out regulation and dredging at some river sections of Wusongkou and Huangpu River, which allowed 10,000-ton ships to enter Huangpu River by the tide and met the requirements of ship development and economic development at that time. In the 1930s, Shanghai port had become the shipping center of the Far East, with an annual cargo throughput as high as 14 million t. Its import tonnage of ships ranked seventh in the world, making Shanghai an important port city in the world.

After the founding of New China, five port areas were developed, including Zhang Huabang, Jungong Road, Gongqing Road, Zhujiamen, and Longwu. Baoshan, Luojing, and Waigaoqiao port areas were also built on the south bank of the Yangtze River estuary. In addition, Baosteel Group, Shidongkou Power Plant, and Waigaoqiao Power Plant also built their own special terminals. As a result, the throughput capacity of the Port of Shanghai has constantly been increasing, which plays an important role in advancing the city development of Shanghai and the economic development of the Yangtze River basin and the whole country.

On December 10, 2005, the first phase project of Yangshan Deepwater Port was completed and put into operation. The Yangshan Bonded Port opened at the same time, marking an important milestone in the development of the Shanghai International Shipping Center. By the end of 2006, the Port of Shanghai Port has 1,140 berths and terminals of various types, including 171 cargo handling berths of over 10,000 tons, with the waterfront 916 km long in total. Its inland river port area has 818 berths, with a maximum docking capacity of 3,000 t.

By the end of 2016, the Port of Shanghai established container cargo trade with more than 500 ports in 214 countries and regions around the world and operated more than 80 international routes. In 2020, the annual container throughput of the Port of Shanghai reached 43.5 million TEU, ranking first in the world for 11 consecutive years.